# Thought, Law, Rights and Action in the Age of Environmental Crisis

# Thought, Law, Rights and Action in the Age of Environmental Crisis

*Edited by*

Anna Grear

*Reader in Law, Cardiff Law School, UK; Director, Global Network for the Study of Human Rights and the Environment (GNHRE); Editor in Chief,* Journal of Human Rights and the Environment

Evadne Grant

*Associate Head of Department of Law, University of the West of England, UK; Coordinator, Global Network for the Study of Human Rights and the Environment (GNHRE); Editor,* Journal of Human Rights and the Environment

Edward Elgar PUBLISHING

Cheltenham, UK • Northampton, MA, USA

Published by
Edward Elgar Publishing Limited
The Lypiatts
15 Lansdown Road
Cheltenham
Glos GL50 2JA
UK

Edward Elgar Publishing, Inc.
William Pratt House
9 Dewey Court
Northampton
Massachusetts 01060
USA

A catalogue record for this book
is available from the British Library

Library of Congress Control Number: 2015933457

This book is available electronically in the **Elgar**online
Law subject collection
DOI 10.4337/9781784711337

MIX
Paper from
responsible sources
FSC® C013056

ISBN 978 1 78471 132 0 (cased)
ISBN 978 1 78471 133 7 (eBook)

Typeset by Columns Design XML Ltd, Reading
Printed and bound in Great Britain by T.J. International Ltd, Padstow

# Contents

# Contributors

**David Bollier** Activist, writer and policy strategist; Co-founder of the Commons Strategies Group; Senior Fellow at the Norman Lear Center at the USC Annenberg School for Communication and writes technology-related reports for the Aspen Institute.

**Lorraine Code** Distinguished Research Professor Emerita of Philosophy at York University in Toronto, Canada, and a Fellow of the Royal Society of Canada.

**Sean Coyle** Professor of English Laws, Birmingham Law School, UK.

**Kate Donald** Director of the Human Rights in Development program area, Center for Economic and Social Rights in New York, USA.

**Gitanjali N Gill** Reader in Law, Northumbria University, UK.

**Evadne Grant** Associate Head of Department, Bristol Law School, UK; Coordinator, Global Network for the Study of Human Rights and the Environment.

**Anna Grear** Reader in Law, Cardiff University, UK; Director, Global Network for the Study of Human Rights and the Environment.

**Tom Kerns** Professor of Philosophy, North Seattle College, USA; Executive Director, Environment and Human Rights Advisory.

**Andreas Philippopoulos-Mihalopoulos** Professor of Law and Theory, University of Westminster, UK.

**Margherita Pieraccini** Lecturer in Law, University of Bristol, UK.

**Burns H Weston** Bessie Dutton Murray Distinguished Professor of Law Emeritus and Senior Scholar, University of Iowa Center for Human Rights, USA.

# Table of cases

## AFRICAN COMMISSION AND COURT OF HUMAN AND PEOPLES' RIGHTS



## EUROPEAN COURT OF HUMAN RIGHTS

## INTER-AMERICAN COMMISSION AND COURT OF HUMAN RIGHTS



## INDIA

## UNITED KINGDOM



## OTHER CASES

# Introduction: thought, law, rights and action in an age of environmental crisis – in search of better future histories*

**Anna Grear and Evadne Grant**

The original GNHRE Oñati International Workshop 2012 – by which this edited collection is inspired, from which it draws and to which it adds – was set up in full knowledge that the relationship between 'human rights' and 'the environment' (both as constructed (legal) domains) remains both complex and fractious.

At the institutional level these two domains (even taken as unproblematic referents to commonly understood concerns) still have an uneasy connection notwithstanding increasing evidence of their interconnections, while at the philosophical level it has often been asserted that the individualism of human rights makes them inherently ill-suited, as a moral and juridical category, to address the inherently more collective – and less 'anthropocentric' – concerns of environmentalism.

Human rights and environmental law are also unevenly theorised. Human rights scholarship is richly reflexive, manifesting an intense, sustained and energetic degree of contestation and theoretical disputation. It is multi-faceted and multi-perspectival: its traditional accounts and

* This edited collection was inspired by a series of articles resulting from a workshop, 'Human Rights and the Environment: In Search of a New Relationship', held at the International Institute for the Sociology of Law, Oñati, Spain, 14–15 June 2012 and coordinated by Anna Grear (Cardiff University). The editors of this collection would like to thank Karen Morrow, Louis Kotze and Andreas Philippopoulos-Mihalopoulos for their comments on an earlier version of this introductory chapter – and to thank, in particular, Lorraine Code for her close, analytical engagement with that earlier version; the anonymous reviewers for the Oñati Socio-Legal Series and Sol Picciotto, Cristina Ruiz and Angela Melville for their comments on and support for the original articles by which this collection was inspired.

assumptions are routinely interrogated by critical, subaltern legal scholarship. Meanwhile, despite a growing interest in critiquing so-called 'anthropocentric' perspectives, environmental law remains largely Eurocentric, dominated – in the main – by Western cultural assumptions and practices of law and governance – while much environmental law scholarship focuses primarily upon on a wide range of legal and regulatory responses to environmental pressures.

The multi-layered complexity of the relationships between human rights and environmental law; between human rights and environmentalism and between humanity and the living order suggests the need to explore the dynamics between philosophy (including jurisprudence), legal doctrine and structures, and the more fluid realms of the socio-ecological, reflected by, among other things, community-led praxis.

The workshop, *Human Rights and the Environment: In Search of a New Relationship*, sought to bring together a group of scholars, lawyers, policy analysts and others to work together – through wide-ranging, careful and at times uncomfortable conversation – on re-imagining the relationship between human rights and the environment. The framework of the event was explicitly designed to move from philosophical foundations, through consideration of current legal responses and frameworks, and onwards to embrace experience-led insights emerging from embedded social activism and NGO engagements. This edited collection broadly traces the organisational structure of the 2012 workshop, focusing on four themes: 1 Philosophical Investigations; 2 Reconfiguring the Legal; 3 Activism and Praxis and 4 Multi-level Reformulations.

## PART 1: PHILOSOPHICAL INVESTIGATIONS

This section of the collection opens with Lorraine Code's critical reflection upon structures of mastery and their political resilience – revealed by practices of epistemic domination in the context of the climate-change debate. Code addresses the theme of 'Doubt and Denial: Epistemic Responsibility Meets Climate Change Scepticism' with the purpose of making the interconnections between three nodes of epistemic engagement legible. She characterises these nodes as being, first, the fraught and politically freighted rift between ecologists and climate-change sceptics; secondly, issues concerning the evaluation of testimony as indicators of questions concerning subjectivity, agency and freedom – and in particular, the impossibility of the abstract, 'un-located' autonomous gender-neutral subject of Western epistemology; and finally, the 'ongoing ad feminam treatment [of Rachel Carson's work by] the science sceptics'. This third node of engagement, for Code, presents a nexus in which the first two converge and are exposed (through her examination of Carson's treatment) as being sites of the operationalisation of sexed-gendered specificities constructing the epistemic 'authority' of (male) subjects smuggled within and behind the structure of the generic 'objective' epistemic subject of Western mastery. Deploying a single text, *Merchants of Doubt* by Naomi Oreskes and Erik M Conway, Code notes the way in which a false neutrality operates to sustain climate doubt and ignorance in the predominantly white, Western world and to excise the fact that 'specifically situated, and not disinterested, human subjects' produce a 'science' dedicated to climate-change scepticism. This 'science' is committed to the notion – fundamental to the epistemic operationalisation of climate denial – of the homogenous human subject: a bland construction perfectly fitting the screen of 'objectivity' designed to foreclose critique and/or resistance. Code argues that the growing significance of testimony in social epistemology creates spaces for engagement with questions concerning the accepted division of intellectual-epistemic labour in Western societies. This also provides the opportunity to interrogate the related construction of the epistemic power 'to know'.

What Code exposes is, in effect, a politics of *disguised* epistemic *location*. She urges a conscious and resistive divergence from the 'neutral', individualistic mode of knowledge production towards a communal one that openly acknowledges the *limits of our knowing*. Relatedly, she argues for acknowledgment of the 'need for critical analysis of some of the taken-for-granted assumptions about subjectivity that prevail in epistemic communities, and for ascertaining the place of those communities in constructing and circulating public knowledge'. Thus, Code brings the politics of subjectivity into full view, arguing that subjectivities elevated as 'either positive epistemic exemplars or as the reverse' should be subjected to genealogical scrutiny – to examination, in short, of exactly how, and why, the epistemic status of such subjects is given or withheld.

Code's analysis is decidedly feminist. It attacks epistemic mastery by exposing it as an exercise of power. She deliberately characterises situated advocacy as being a feminised practice engaging the adversarial, legalistic (and masculinist) construction of 'authoritative' scientific and legal knowledge. If knowledge is power (the relation implicit in epistemic mastery) then the epistemically responsible, critical, unmasking of 'neutral' knowledge is a thoroughly necessary – and highly political – act. Code invites us, in effect, into a struggle for epistemic justice.

Philippopoulos-Mihalopoulos, while sharing some common ground with Code, moves beyond Code's immediate concern with questions of epistemological mastery (though doubtless he would agree with her critique) to seek out new ontological foundations for a 'critical environmental law' lying beyond, as he would see it, the contemporary fixation with epistemological questions *in the place of* ontological ones. Addressing the theme of 'Actors or Spectators? Vulnerability and Critical Environmental Law', Philippopoulos-Mihalopoulos seeks to explore what it might mean to embrace our 'thrown-ness' (Heidegger's phrase) into a new, category-confounding, vulnerability-exposing open ontology. Philippopoulos-Mihalpoulos argues that the reified ontological polarisation between humans and the 'environment' has been *replaced*, of late, by an epistemological debate between anthropocentricity and ecocentricity. This is a move that he regards as problematic in two key respects. First, it distances us from the ontology of the human/non-human interaction and, in this sense, replaces 'ontology with an epistemology of supposed action'. Secondly, the move legitimates, in his view, the 'perennial problem ... of "the centre"' (a point having considerable political resonance with Code's critique of the dominance of the Western epistemic subject).

For Philippopoulos-Mihalopoulos, the move from ontology is a form of avoidance – a move as he puts it, avoiding exposure to the vulnerability of being in the wild and dizzying space of the 'middle' rather than in the comforting dominance of the epistemic 'centre'. Philippopoulos-Mihalopoulos invites his readers to re-experience their true ontological location: as beings exposed, *thrown-in*, 'abandoned' to the forces of a-centricity that characterise the true complexities and energies of existence. There is a need, he argues, for semantics that 'move away from the glib safety of the centre and [its] consequent, misaddressed dialectics [arguing that][a n]ew semantics ... must be sought on a surface of acentric continuum, traversed by velocities and pauses that do not easily subscribe to centralising generalisations'. In short, this is an invitation to give up epistemic power and privilege. It complements Code's emphasis upon the exposure and resistance to the central epistemic subject's power of 'knowing' – but moves us into a multi-planar dislocation of the 'human' viewpoint.

Philippopoulos-Mihalopoulos follows Deleuze and Guattari onto the 'the plane of immanence' – inviting a fall, as Philippopoulos-Mihalopoulos puts it, into the 'all-embracing sum of folds and falls and connections that contains its own origin, causality and teleology without transcendence'. Ontology and epistemology retain no clearly demarcated binary relationship, for Philippopoulos-Mihalopoulos argues that the plane of immanence 'operates on both an epistemological and an ontological level', allowing no division between the observing spectator and the involved actor, no distinction along the lines of human/natural/artificial/technological – for all these 'fold into each other and constantly emerge as epistemological and ontological hybrids'.

Key to Philippopoulos-Mihalopoulos's argument is the embrace of vulnerability. Vulnerability breaks with the need for any observational distance: vulnerability radically undermines the impossible neutrality that Code's paper so vigorously attacks. Vulnerability is full exposure to the 'space of the middle': '[t]here is no screen to hide behind, no distance afforded by epistemology, no negotiating moments of discourse. The surface is pure ontology, inescapably filled with a brutality of continuous, uninterrupted presence'. This is a vulnerability that invites – in a Heideggerian mode – a full exposure to ontological 'thrownness' in the world.

What might the implications for law be? Philippopoulos-Mihalopoulos offers his readers an account of a critical environmental law that opens out the paradoxical nature of environmental law itself as that field of law potentially most open, most vulnerable, to 'non-legal' inputs and yet simultaneously in need of its own juridical identification and partial closure.

Addressing a related challenge, Margherita Pieraccini examines two representations of the legal relationship between human rights and environmental regulation, also drawing on the work of Heidegger – this time to present alternative foundations. Pieraccini divides her discussion into a reflexive duality: the first part of her chapter is dedicated to an examination of 'the oppositional view' that 'corresponds' to what she calls 'the initial phase of the relationship between human rights and public environmental regulation'. This by necessary implication focuses on a profound subject-object division between human rights and 'the environment'. In law, this is expressed as on the one hand, a subjective rights-based approach circling around a human subject and on the other hand, the 'external, state regulation' of a 'de-subjectivised good: the natural environment'. For Pieraccini, the assumption lying at the basis of this view is the Cartesian-Kantian subject-object split. However, the Kantian subject-object split, in particular, especially because of the influence of the Kantian moral and legal subject on human rights law, invites a direct critical engagement with the topic of epistemic mastery as it has been worked out through law: the Kantian subject-object duality, Pieraccini suggests, 'positions human rights and environmental regulation in two bounded and conflictual legal realms'.

Pieraccini's response to this is to analyse what she calls the 'interactive view'. This approach, her argument implies, is increasingly discernible in contemporary legal and policy developments and in academic scholarship. She argues that there is a 'greening' of human rights law and a related 'righting' (as she puts it in verb form) of environmental law. This dual movement reshapes, Pieraccini suggests, the boundaries and connections between the two fields. The weakness, however, of the interactive view is that it is an empiricist, policy-dominated phenomenon in need of a more radical critical engagement with traditional ontological assumptions. The gap between 'human subjects' and 'the environment' – the subject-object dichotomy lying behind, as Code would insist, the epistemology of mastery – remains intact.

Pieraccini is interested in exploring how the abandonment of Cartesian-Kantian dualism and a 'focus on the question of the meaning of Being' enables relational thinking. For an ontological account capable of producing such a renewing imaginary, she turns, naturally enough perhaps, to Heidegger.

Heidegger, she suggests, was the first philosopher to re-open 'the problem of Being as distinguished from beings'. A Heideggerian account suggests that 'the two separate legal spheres of human rights and environmental regulation are parts of, and constituent of, a particular way of revealing Being that conceals practices of dwelling and ethics of care'. There is thus a need to move beyond existing concepts to open up spaces where a 'poietic' (in the Heideggarian sense of 'bringing forth') view of law can be disclosed. This argument, as will become clear later, has considerable resonance with the phenomenology of the commons-based dynamics discussed by Burns Weston and David Bollier in their chapter in this collection.

Code, Philippopoulos-Mihalopoulos and Pieraccini share some fundamental concerns and philosophical intuitions, for all the nuanced differences in their accounts. The selves that emerge from their accounts, taken together, are situated selves – not omniscient or radically separate from 'nature', but nodes of emergence within a systemically complex, vulnerable and contingent set of interwoven actualities. Code's responsibilised epistemic subject is humble – self aware as to the politics of location – and positioned in tension with the epistemic subject assuming the centrality of mastery. Philippopoulos-Mihalopoulos 'throws' an epistemically limited subject into a radical, open ontology in which easily assumed distinctions simply break down and for which the exposed vulnerability of the middle is the antithesis of 'the centre'. Pieracinni, while she retains more ontological duality in her understanding than Philippopoulos-Mihalopoulos, invites a new sense of situtatedness in which law itself, as a poeitic dynamic, might emerge into new, interwoven relationalities transcending traditional dyadic structures.

In 'Radicalism and Conservativism in Environmental Law', Coyle argues from a natural law position that re-inscribes precisely the kind of human uniqueness that Philippopoulos-Mihalopoulos, at least, would be keen to cast aside – a uniqueness, for Coyle, hinging on the nature of the human being 'as a rational animal'. Indeed, Coyle argues that what Western philosophical accounts have in common is the central idea of the human being as being 'in certain respects, *apart from nature*' – and that for him – this 'apartness' has no intrinsic sense of 'aboveness' or of superiority. It is, however, central to ethical understanding.

For Coyle, 'the separation of Mankind from the rest of nature is … as important as its situation within a wider order of nature', and hinges on the human facility for language – a facility that means that 'for human beings … there is not "the world" and then "an interpretation" of it … there is no separation between our experience of the world and the meaning we give to it'. The implication of this is that 'there exists no platform from which we are able to abandon a "human" view of the world and adopt a perspective from which human concerns, interests and structures of meaning and understanding are not central'. Accordingly, radical ecological and postmodern viewpoints that insist on taking the 'standpoint of the other' mistakenly over-stretch a human capacity for something more like a standpoint empathy:

> Radical 'ecocentric' thought began from the assumption that environmental concerns must be derived from an ethical standpoint that requires us to give up a distinctly human perspective. Such a standpoint is not available. What in fact is required is a standpoint from which ethics does not terminate in human concerns, but is properly transcendent of them.

This 'proper' transcendence is found in systems of thought (such as natural law, Christianity, Daoism, Confucianism) that yield an environmental ethic that separates intrinsic environmental value from human concerns and purposes – a task not assisted by the assumption of an ecologically radical standpoint assuming an impossible epistemic transcendence. In this insistence, at least, Coyle's account echoes that of Code, Philippopoulos-Mihalopoulos and Pieraccini – each of whom, from very different philosophical starting points, insist upon epistemic situatedness and an accompanying form of epistemic humility very far removed from a transcendence of 'humanity'.

Coyle's central concern is to trace out the tensions between what he calls 'anti-conservative' 'radical' aspects of environmental legal thought and the deeper, 'conservative' (conserving) ideas of environmental law (and human rights). Law, he argues, functions as that which reflects and protects bonds uniting society, and modern environmental principles draw upon the deepest aspects of property law and tort as mechanisms of careful social-interest balancing. Such conserving aspects of legal relations are abandoned at the risk of distorting the legal order such that it becomes a mere 'system of adventitious adjudication between competing instrumentalities'.

For Coyle, environmental legal radicalism – such as that emerging from feminist, eco-feminist, Marxist and other critical positions – reflects 'anti-conservative forces within politics which, rather than seeking the deeper integration of a form of social life, go off in search of an ideal re-ordering of society; a re-ordering which is nevertheless understood in so many different and non-compossible ways'. Into this tension, Coyle inserts the contribution of human rights – which despite the fact that they, like environmental law, reflect conservative and anti-conservative elements – offer a more direct mode of expression for conservative (in Coyle's intended sense) ideas. This is because human rights point towards the important fact that 'certain common conditions are required if [human beings] are to flourish in this world, and that such flourishing can therefore be achieved or promoted only as a common endeavour … A clean and habitable environment is certainly one of these commonalities'. Crucially, however, this commonality emerges from a richer understanding of society than is dominant in the modern liberal democratic state, in which humans are 'consumers' competing to achieve diverse individualistic goals. For Coyle:

> the doctrine of human rights promises, though no more than partially, to return us to something closer to the pre-modern conception of the person: as a being with a unique and individual soul, for which he alone is responsible, but which is embodied in a fleshly home whose essential characteristics and problems, and whose ultimate well-being, are inexorably linked with others of his kind. Human rights emphasize the collective context in which human life runs its course. In so doing, they may perhaps produce a more integrated understanding of law; and of rights that are ultimately tempered by the needs of a common environment.

Albeit by a radically different route, Coyle returns us to the ethical intimations of materiality, embeddedness and situatedness. However, this emphatically does not reflect the immanent open ontology of Philippopoulos-Mihalopoulos, nor that implied by the ecological epistemological project of Code. Nor is it Heideggarian. Coyle's account reflects an ontology of the human self as a duality of body and soul – a being not completely immanent – albeit that this human's fallibility of perspective, embodied limitation, social situatedness importantly underlines a human incapacity to escape the limitations of human perspective itself.

What emerges from the convergence of all four of these chapters is a productive provocation to tease out some related and challenging puzzles. What standpoint positions are even possible? What kinds of epistemic empathy and feats of ontological imagination open up future histories that simultaneously exceed and engage critically with the practices and patterns of power so central to the destruction driving our climate crisis and related ecological crises? What concept of 'nature' (if any) might be liberatory – or oppressive – and for what/whom? Hidden within all these questions, moreover, is the substrate that Code's work relentlessly forces us to confront – the central question of the political and epistemic constitution of forms of subjectivity – and the forms and processes of their construction and dismantling: Who and what – and how – is privileged by the way we construct our philosophical, juridical, political, scientific, social, economic and ecological imaginaries?

## PART 2: RECONFIGURING THE LEGAL

We turn now to law more directly.

In 'Human Rights and Environmental Protection in India: The Judicial Journey from Public Interest Litigation to the National Green Tribunal', Gill explores the fractious pressures between socio-economic development and environmental protection in an analysis of the role of the Indian judiciary in developing a jurisprudence responsive to both human need and environmental degradation.

The Indian context, Gill suggests, presents a paradox produced by the 'convergence of the right to environment and the right to development'. India's vast population – the sheer pressure of human numbers – along with rampant poverty, inflation and corruption – present particularly pressing challenges in response to which India embarked on a 'new path of economic growth'. This has not, thus far, produced the collective benefits that might be aspired to, but has, instead, led to a widening gap between the rich and poor in India. Meanwhile, the growth agenda also frequently conflicts with human rights values and with the protection of the environment.

Gill's chapter addresses this paradox, and sets out to trace 'the successes and failures' of public interest litigation 'as a legal procedure to … promote economic activity and simultaneously to maintain the environment', culminating in an analysis of the National Green Tribunal. The example of India presents a fascinating and salutary example of re-inventing the legal: in particular, as Gill's analysis establishes, the Indian judiciary have responded to poor regulatory responses to environmental degradation by forging 'innovative and creative [modes for the protection of] ecology and the environment'. Significantly, Indian public interest litigation opens a 'broad-based, people-oriented approach that provides access to justice and has become a "wheel of transformation" for victims of environmental degradation'. However, weaknesses and shortcomings vexing the public interest litigation procedures meant that further innovation and change were required. The Indian judiciary, again, rose to the challenge, delivering three judgments at the constitutional level that placed normative pressure on the Indian government to respond. Eventually, the trajectory of development culminated in the passing of the National Green Tribunal Act, in June 2010.

The Indian National Green Tribunal is an impressive model for future re-configurations of legal process, revealing a sensitive juridical response to environmental complexity. The decision makers on the Tribunal hold expert interdisciplinary qualifications, and 'judicial and scientific bench members act as joint decision makers of equal standing'. Here, arguably, are traces of a juridical strategy reflecting ample space for a critical environmental law theorisation – responsive, perhaps, to the quintessential openness of environmental law that Philippopoulos-Mihalopoulos's chapter introduces.

The interweaving of the right to life and to a healthy environment in the Indian constitutional jurisprudence, likewise, emphasises the materiality and the vulnerability so central to questions of human and environmental justice. As is the case all over the world, in India the most pressurised points of convergence between interests occurs in conflict between the baseline, material, corporeal and situated needs of the 'poor' – in this case – the 'tribal poor' – and the 'powerful rich'. In response to this highly uneven dichotomy – so representative of wider patterns of the contemporary juridical order – the Indian judiciary offer a rights-sensitive response. Gill suggests that the National Green Tribunal is 'laudable as its judgments recognize that "development" is the essence of any pragmatic and progressive society based upon a "balancing act" of not only supporting economic benefits but also encompassing the guarantee of protecting the environment and the fundamental human rights of the poor'.

Placing the rights and interests of the most marginalised members of society in a direct and intimate proximity to environmental concerns reflects, arguably, a critically attuned judicial response to climate injustice and to other recognisable patterns of oppression that bring non-dominant humans, animals and ecosystems into an intimate proximity within the titled order of neoliberal capitalism. It is no surprise, then, that the National Green Tribunal's efforts are being 'hampered by' the Indian government's economic policies in the pursuit of 'unbridled growth':

> Increasing political pressure is being exerted to promote economic development via public and private investment at the cost of the environment. The government is planning to amend the forest laws to increase economic activities in states, which hold some of India's most productive forests and also the majority of the tribal population. The government also seeks to dilute the Forest Rights Act that requires the consent of the local tribal population concerning change of use of forest land particularly mining for minerals ... [and by various means] excluding project-affected tribal people from the final decision-making process ... It is hoped that the [National Green Tribunal] will hold fast to its basic and established jurisprudence in the face of these new economic policies.

Human rights are often attacked for their individualism as well as for the disconnections evident between categories of rights. Evadne Grant, in 'Re-imagining Adjudication: Human Rights Courts and the Environment' examines this critique, conducting a review of the approaches of three regional human rights courts to reveal an iterative interpretive construction of an expanding relationship between human rights and the environment. Arguably, the central theme of Grant's account circles around the relatively prosaic, but powerful, generation of new sets of interconnections – between rights and rights, between rights and communities, and between rights, communities and environment.

Grant concedes the force in critical readings of the individualism and disconnectedness of human rights. She broadly accepts the fact that human rights approaches to environmental protection face important conceptual and procedural constraints, but argues that a close analysis of the developing jurisprudence of the various regional courts and related human rights institutions (in this case in the European, Inter-American and African regional systems) shows promising signs of an expanding jurisprudence pointing in another direction. There is nothing inherent in human rights law, her analysis suggests, that means that human rights reasoning cannot embrace group and community interests extending important and meaningful protection to the environmental contexts in which human well-being – whether individual or collective – is unavoidably situated.

Grant's analysis suggests that notwithstanding manifold institutional and textual constraints, the jurisprudence of the regional courts reveals a quietly burgeoning potential for change effected through skilful interpretive extensions of traditional rights-based reasoning. Not only are courts connecting substantive human rights with each other in new ways, implying important community-dimensions of their meaning, but there is, in particular, a relatively un-trumpeted revolution being enacted by the quiet, constrained radicalism of procedural environmental rights. Grant's analysis suggests that despite their somewhat technical surface, procedural environmental rights are an increasingly important locus of immensely valuable modes of community inclusion. Such rights, she suggests, clearly recognise that environmental hazards affect whole communities. Despite varying modes of implementation across the three regional human rights systems, and despite the fact that procedural rights might lend themselves to State avoidance of more substantive standards, the quiet revolution enacted by the interpretation of procedural rights richly suggests the innovative, value-led determination of the regional courts to extend meaningful environmental protection to human beings and communities.

In Grant's analysis perhaps the most promising development concerns the increasing level of normative interweaving visible in the increasing tendency of judges to reach across jurisdictional boundaries to draw inspiration from the jurisprudence of other regions and even from national constitutional orders. Indeed, such promising judicial cross-pollinations seem to have the effect of drawing the various international and regional treaties into ever more intimate conversational proximity with national constitutional provisions – suggesting a degree of normative convergence locating human rights within broader, ecological contexts (ecologies both 'natural' and juridical).

Ultimately, Grant concludes, while many might argue for a radical sweeping aside of existing paradigms, the incrementalism of the quiet revolution of judicial approaches points the way towards a new paradigm already being tested and implemented. Importantly, the normative foundations for this development are *inherent* to human rights law. Grant argues that from this perspective, while criticism of human rights individualism and disconnection is important, it may overlook the more prosaic realities of an environmental human rights law whose successful emergence is marked precisely by its lack of a radical break with the status quo – and which is, by implication at least, the more effective for that.

## PART 3: ACTIVISM AND PRAXIS

Kate Donald's chapter 'Human Rights Practice: A Means to Environmental Ends?' implicitly addresses doubts concerning whether or not human rights practice in its current dominant forms can tackle the challenge of climate change and global environmental degradation. Donald argues that while the symbiosis between human rights and the environment is increasingly well established, human rights – for all their potential to contribute in concrete ways to moving an environmentally responsible agenda forward – have not thus far realised it. Whether they can or not still remains a relatively open question. In fact, Donald's analysis implies that human rights cannot assist in the search for environmental justice or fuller environmental accountability unless we adopt and pursue 'radical or hybrid approaches, with a view to articulating a strategy for activism and praxis that can capture the real and lived inter-connectedness of human rights enjoyment and environmental factors more meaningfully'.

In order to defend this thesis, Donald addresses three dominant existing approaches to human rights praxis – each of which reflects different understandings of the relationship between human rights and the environment. First, there is the strategy of lobbying for the adoption and institutionalisation of new environmental human rights. Secondly, there is the strategy of deploying existing human rights mechanisms to tackle environmental harms. Thirdly, there is the strategy of taking a human rights-based approach to environmental practice.

Donald concludes that the first strategy, while it holds out hope for activists, is unlikely to offer much in more substantive, juridical terms due to a range of institutional and political weaknesses in the dominant formulation of human rights. There are, moreover, she suggests, fundamental questions as yet unanswered by the strategy – particularly concerning its location in the unproductive dichotomisation between anthropological and ecological approaches. The current challenges reflect the need to embrace both the environmental and human costs of failing to respect the Earth system – but Donald remains doubtful that even a human rights treaty with universal applicability could overcome the concerns represented by the intractable questions presented by the controversies surrounding the relationship between human rights and the environment. The second strategy – of using existing human rights mechanisms (courts, UN treaty bodies, special procedures) to protect the environment – Donald considers to be 'worthwhile … in certain cases [but to have] limited capacity and reach' – noting that while environmental issues are often united with social and economic rights, they suffer from the same weaknesses concerning their perceived justiciability status. This is a key weakness, arguably, in strategies seeking to deploy existing socio-economic human rights entitlements as an avenue for the protection of environmental rights. The third strategy, the deployment of a 'human rights-based approach' to environmental matters, is more promising – but particularly 'when applied to activism and action on the ground with a sensitivity to context'.

This third strategy arguably reflects a socio-legal critical approach – one moving into forms of eco-humanitarian grassroots organisation and engagement and implying the energies emerging from responses to localised, environmental injustice, and also one highly connected, we would suggest, to Code's argument for advocacy conducted in the full light of detailed, situated analysis and sensitive to situated subjectivities.

A rights-based approach, in Donald's analysis, and understood in the way she defines it, has the potential to be subtle, flexible and responsive to structural and multi-faceted causes of rights violations and lack of rights enjoyment. Donald gives specific core content to the third broad strategy: her emphasis is upon accountability; equality and non-discrimination; participation and empowerment. Each of these terms, she notes, has a particular meaning in human rights contexts. These meanings, it should be noted, directly address the problem of marginalised subjectivities – and implicitly critique the dominant hegemonic subject – because they imply a notion of equality given substantive content by gender-sensitivity, non-discrimination and an implication of affirmative action. Thus, while Donald emphasises the significant potential for rights-based approaches to 'address and question structural and systemic causes of disempowerment and vulnerability to (for example) poverty or environmental harm', it is of note that this relies upon stepping back from law's institutional provision to operationalise more fluid approaches and adaptive, ground-level strategies. This less predictable way of working holds out hope of facilitating a more reflexive, liberated and effective form of human-rights-based practice.

Such approaches, however, need to keep a clear sense of the complexity of human rights as semantic, semiotic and juridical tools. It seems that what such rights-based approaches have the greatest potential to do is to place the battle for human rights meanings at a more localised, contextualised level characterised by a certain productive contingency and flexibility of response. This is to be welcomed; yet there remains a need for explicit caution concerning the assumptions to be held about human rights – a view evidently shared by Donald, who reserves her greatest enthusiasm, in the end, for a fourth, alternative '"principled strategy" for the embedding of the interdependence of human rights and the environment in human rights practice, based on a realistic interpretation of the strengths and weaknesses of human rights in its various shifting and contested forms'. Donald's is, then, a decidedly non-monolithic reading of human rights – one placing their haunting ambiguities firmly in view and entirely open to a creative blend of the strengths of various strategies – while aware of their weaknesses – and blurring the boundaries, potentially, between rights-based activisms and other social movement claims for social and climate justice.

Indeed, it is as *claims* that human rights may do their deepest work. For Donald, 'claims-making is a distinctive part of human rights practice and has intrinsic worth and radical potential'. Claims-making on this view forms a kind of transformative advocacy close to the advocacy that Code champions in her chapter. This is the kind of claim-making, narrative-telling, argument-making intervention that produces shifts in consciousness precisely by problematising what is taken for knowledge in the fraught environmental justice context. This makes a distinct gesture away from law to the discursive power of human rights as an *idea* animating activist resistance. This is a messy, energetic conception of rights-claims that allows us – as Baxi puts it (cited by Donald) – to '[construct] some new alternate futures beyond the new paradigm of trade-related, market friendly and environmentally hostile human rights'. Indeed, the struggle for environmental justice, for Donald, also 'has the potential to contribute to a much-needed task: the renewal and reconstruction of human rights in a more radical form'. Thus, in the way that human rights, for Coyle, point to a bond-conserving sensibility capable of addressing the separative fantasies of the modern liberal consumer state and directly assist environmental law in its 'conserving' role, for Donald, human rights themselves become transformed by social movement energies directed by the centrality of the environment to questions of justice. Against a background in which, despite the growing importance of the environment in human rights discourse, human rights are increasingly marginalised at the institutional level within international environmental fora, Donald argues that 'those who are concerned with environmental protection and with human flourishing in a healthy environment should encourage, use or welcome a "principled strategic" use of human rights, based on an analysis of what can best realise inevitably contextual visions of justice, dignity and flourishing'. In particular, this re-energisation of human rights becomes all the more urgent in the context of the emergence of capitalism's reinvention of ecological justice in the form of the green economy – a concept 'lacking even a human focus'. Human rights practice, by contrast, holds out hope for a 'principled yet strategic, profoundly context-dependent' approach that accepts shifting power relations as core to its analysis. Key to this – at all times – are the energies of social movement actors on the ground.

Tom Kerns would undoubtedly wholeheartedly endorse the importance of this kind of ground-level claim-energy. His careful analysis, in 'Schopenhauer's *Mitleid*, Environmental Outrage and Human Rights', drives directly at the activist's concern with the energies emerging from the lived-experience of violation, of environmental and human degradation, and the 'outrage' energising human rights in their most liberatory mode. Kerns's analysis is intrinsically context-responsive, located and situated. It suggests not only the ultimate ontological unity between humans and environment but the productive, visceral intimacy between outrage, speech and the search for eco-humane justice.

Kerns argues that suffering that results from exposure to environmental violations evokes a sense of moral outrage – and that this outrage can be grounded in Schopenhauer's important philosophical work on compassion. Human rights, argues Kerns, function as an important source and form of validation for outrage. Importantly, Kerns also argues that Schopenhauer's account leads towards an explanatory grounding for the 'long-recognized importance of personal narratives in human rights work'. In short, human rights confirm and publicly validate what compassion first intuits – an insight leading to three practical implications for environmental activism: the central 'importance of personal narratives detailing the direct impacts that environmental assaults have caused'; 'the practical value of formal, detailed human rights assessment reports specified to a given situation'; and 'the value of community-led public inquiries, such as the 2006 People's Inquiry in New Zealand and the 2011 Permanent People's Tribunal in India'.

Kerns's central philosophical claim is that Schopenhauer's ethic of compassion, his *Mitleids-Moral*, provides a better account of the human experience of moral outrage than do the other systems of ethical thought (Aristotelian, Kantian, Millian) often called upon in human rights philosophy. Kerns's analysis drives at the visceral sense of empathy lying in the radical, critical foundations of human rights when they first emerge and before they are captured, as it were, by human rights law:

> As Lynn Hunt says in *Inventing Human Rights: A History*, 'we are most certain that a human right is at issue when we feel horrified by its violation'. When we see or hear of such horrible injustices we are often seized with shock and outrage and blurt out, if only to ourselves: 'How awful! How could anyone do such a thing?'

This is the response of compassion – a compassion that for Schopenhauer (and Kerns) is 'participation in' another's suffering.

The mystery of this compassion is explained by an empathic openness born of the ontological reality of our fundamental unity of being: 'All these multiple beings in the world of phenomena, i.e., all the temporal-spatial multiplicities of our experience, are, for Schopenhauer, phenomenal expressions of the same underlying fundamental reality, *Der Wille*': We are all different manifestations of the same underlying reality. It is this ontological fact that provides the metaphysical explanation of our exposure to the sufferings of another.

This explanation reticulates with the idea that we can feel the sufferings of others resonate in our own embodied vulnerable being. For Schopenhauer, as Kerns makes clear, our fundamental unity with all things means that we experience the sufferings of others as if they were our own. This fundamental insistence upon a kind of empathic corporeal communication growing from ontological commonality is, in a sense, a rather different way of driving at the vulnerability central to the structure of existence. It intimates, without conforming with, the sense of the vulnerable space of the middle that Philippopoulos-Mihalopoulos emphasises in his contribution to this collection, suggestively hinting at the radical 'suchness' of what exists as being a fundamentally vulnerable, open field of dynamic movement, shifts and interactions between 'actants' (to use Philippopoulos-Mihalopoulos's term).

Kerns suggests that Schopenhauer's ethic of compassion captures something of the personal, existential sense of outrage at unnecessary suffering. He argues that the international human rights tradition 'provides another level of validation for that sense of outrage and the felt necessity of acting on it'. Human rights norms take the hot energy of compassion and channel it through public structures – enabling recognition of what begins as a most personally felt violation. Human rights, in this sense, appear in Kerns's vision to function as a mode of enunciation or articulation – as a channel for engagement concerning environmental (and human) violation. From this, for Kerns, certain insights follow for environmental justice activism. He suggests that:

> If the central claims of this paper have validity, and if Lynn Hunt's historical account of the role of empathy in the genesis of rights discourse is correct and she is right to argue that 'rights are best defended in the end by the feelings, convictions, and actions of multitudes of individuals, who demand responses that accord with their inner sense of outrage', then at least three practical implications follow for environmental activism.

First, *telling the story*: Kerns emphasises the power of the meaningful personal account of direct impacts and the importance of collecting and documenting these narratives of violation. Secondly, *claiming moral authority*:

> Providing activists with well-researched Human Rights Impact Assessments of their specific situation can help clarify the moral dimensions and values at stake as well as identifying potential pressure points. Foregrounding and documenting human rights standards that apply in that particular situation can publicly validate the felt sense of injustice and legitimise the outrage experienced by those whose lives have been negatively impacted.

Thirdly, and finally, *exercising moral power*: Here Kerns emphasises the potency of community-led public inquiries – pointing to successful examples of community-initiated people's tribunals, which have the advantage of being able to be truly inclusive. These are, in a sense, modes of hearing pain and outrage speak truth to power – a very practical endorsement of Code's powerful advocacy-based argument for ensuring epistemic openness in the face of tactical diminutions of the 'authority' to know, and of Donald's arguments concerning the role of flexible, rights-based approaches. Kerns's emphasis upon *Mitleid* also resonates with a concern for the most marginalised and socio-economically disempowered communities.

The intransigent tensions between situated, embodied, eco-material needs, socio-economically problematic imperatives of neoliberal economic expansionism, activist aspirations and the internal contradictions of rights as paradoxical forms of complicity and resistance adds to the general sense, communicated in the contributions to this collection thus far, of profound urgency of a shared, albeit diversely constructed, search for a renewed responsiveness to contemporary crises. The final two chapters in this collection offer thoughts towards a 'multi-level reformulation'. The contributions in this final section share much in common with some of the earlier contributions, but overtly trace the contours of future-facing developments intimating the shapes of new juridical imaginaries explicitly responsive to socio-ecological complexity and to the need for new modes of responsiveness.

## PART 4: MULTILEVEL REFORMULATIONS

In 'Reimagining Ecological Governance Through a Rediscovery of the Commons', David Bollier and Burns Weston draw upon their earlier work to 'argue that the Commons paradigm, especially when guided by human rights law and policy, can help us imagine and implement a new vision of provisioning and democratic governance'. Importantly, Bollier and Weston do not 'propose an abrupt, revolutionary change, but one that can feasibly evolve within the fragile, deteriorating edifice of existing institutions'. For Bollier and Weston, the unique strength of the commons, both as a discourse and as a 'set of ethical social practices' is its combination of a coherent critique of the status quo, its embrace of legal principles with powerful, imaginative implications for the design of proactive, change-making strategies and – last but certainly not least – a 'transnational movement of commoners who are co-creating innovative new provisioning and governance systems that work'. Importantly, especially given the multiply-manifested crises attending the 21st-century situation, the commons has a 'secret potency' that: 'can help to disrupt the State/Market's pathological dependence on economic growth and begin to set enforceable limits on the overexploitation and abuse of Earth's natural systems ("the tragedy of the market")'.

Bollier and Weston make explicit the fact that a re-imagined (and re-experienced) sense of humanity lies at the heart of the power of the commons. The commons is an incontrovertibly social phenomenon that refutes the assumption that human beings are self-centred, value-maximising, rationalistic individuals. The commons, Bollier and Weston emphasise, 'argues that human beings [also] have many more complex propensities – for cooperation, creative initiative, sacrifice for collective interests, idealism, etc'. Commons are, archetypally, the 'intermediate communities' that Coyle writes of. They reflect, arguably, the interwoven self lying at the heart of an ecological epistemological project – as well as the central importance of situatedness in the generation of normative ethical impulses. In a sense, it is precisely their sensitivity to a ground-level politics of location that lends commons-based governance its incredible flexibility and reflexive openness: the greatest power of the commons lies precisely in its 'highly particularistic lived realities'.

The commons framework, as Bollier and Weston make clear, draws upon energetic commons-based scholarship, and upon a 'global movement of commoners that began to emerge in the late 1990s and early 2000s, much of it fuelled by the World Wide Web and free/open source software'. The sheer variety of commons is almost dizzying. At local, national and transnational levels commons proliferate in a kaleidoscopic range of resources. Bollier and Weston name: 'subsistence agriculture, stewardship of water, the re-localisation of agriculture and other production, open access scholarly publishing, new forms of "open source design and manufacturing" for cars, furniture, farm equipment, policies for a "shareable city", scientific data sharing, "collaborative consumption", cooperatives and alternative currencies, among many others'. The focus is on building, ground up, workable modes of 'fair, inclusive and sustainable ways' of 'managing and preserving shared resources'.

For Bollier and Weston, the felt-experience or sense of vulnerability is a powerful explanation for the growing appeal of commons-based self-organisation. It is, in a direct sense, a profoundly human response to the dehumanising impersonal injustice of the 'State/Market' duopoly. Referring to their book, *Green Governance*, Bollier and Weston argue for a 'commons imaginary' that 'calls for innovative uses of State and human rights law and policy to facilitate the formation and maintenance of ecological commons'. In short, they propose that 'modern law should recognise the age-old paradigm of the Commons and provide affirmative recognition and support for the right to common'.

This call, in a sense, demands the bringing forth of a fully ecological epistemic and ethical framework in law and governance structures – a space of creation for a far more open socio-juridical ontology to emerge in tangible, interwoven community-based praxis. The Commons paradigm has much in common – therefore – with other contributions in this collection. It also expands the notion of law itself. Bollier and Weston specifically call for a notion of 'vernacular law', 'vernacular' being a word that (here they quote Illich) 'implies "rootedness" and "abode" and derives from the Latin word *Vernaculum* that described "sustenance derived from reciprocity patterns embedded in every aspect of life, as distinguished from sustenance that comes from exchange or from vertical distribution"'.

For Bollier and Weston, human rights can play a pivotal role in the development of the new paradigm, ensuring the juridical space for this complexity-sensitive mode of grassroots adaptive self-governance to flourish through a 'human right to green governance'. Their contribution openly celebrates the power of grassroots-driven approaches, validated by complexity science demonstrating that 'some of the most robust, stable forms of governance are distributed, self-organised and collaborative'.

It is not too great a stretch to see how this vision returns us to considerations of the dynamism, complexity, vulnerability and energies of an open ontology and to the radical epistemic humility demanded by new philosophical imaginaries drawn from the new materialist visions and critical theories informing, in particular, the work of Philippopoulos-Mihalopoulos and Code in this edition.

Anna Grear's contribution, 'Towards New Legal Futures? In Search of Renewing Foundations' draws from similar inspirations – turning a processual ontology towards an epistemology, ethics and re-imagined set of legal process patterns entirely consistent with the opening chapters of this collection. Grear suggests that the 'animating purpose' of her chapter is to 'reflect upon insights yielded by a lively confluence between critical environmental law scholarship and new posthuman materialist ontologies (from which critical environmental law itself draws)', placing these in 'direct engagement with a search for a renewing socio-juridical imaginary in which the relationship between human rights and the environment might be reimagined'.

Grear contextualises the concerns of her chapter within the emergence of the Anthropocene, reading the Anthropocene construction of a new universal humanity critically, before exploring some broad directions for new human and environmental foundations for law and legal processes. Grear insists that for all the potency of speaking of a common vulnerability in the face of climate change (the level of threat co-situates 'us' 'with countless "other species" and living systems'), the contemporary Anthropocene crisis is characterised by conditions of '*common and uneven* precarity'. What then, is law's response? Grear is clear that the current legal paradigms of human rights and the environment have signally failed to respond adequately. The 'track record' she argues is 'lamentable. International legal regulation and governance regimes exhibit profound limitations impeding genuinely transformative possibilities'. She suggests that:

> If climate precarity – as Chakrabarty argues – 'flashes up' the need for a construct through which to imagine new ways of being and eco-wise future histories, then it will be essential to replace the subject-object relations at the foundation of climate crisis and the unjust patterns of the neoliberal international legal order alike, introducing new, inclusive eco-subjectivities – including new eco-juridical subjectivities.

Fortunately, Grear suggests, the broad outlines of such future-facing possibilities exist. Turning to new materialist philosophy, she draws upon the work of Merleau-Ponty, Barad, Haraway, Philippopoulos-Mihalopoulos, and also of Code, to suggest that a dynamic, new materialist processual ontology is already on offer, one with rich implications for new forms of inter- and intra-species justice – and drawing on the work of Pieraccini and Philippopoulos-Mihalopoulos – able to produce new legal normativities arising from the interactions between non-human animals, receptive humans and living contexts. Additionally, Grear argues, Code's ecological epistemology points the way towards a new ethical sensitivity to the question of situated positionality, making the question of situatedness itself intrinsic to ethical encounter. All of this hinges, Grear's analysis implies, on a rejection of the fundamental subject-object relations underpinning the present system and its construction of legal subjectivity and an embrace of new, dynamic and open interwoven realities in which subject and object emerge as a form of *presencing* contingently 'arrested' for particular purposes.

Finally, Grear turns her analysis towards legal process values, suggesting that not only might law move towards a range of less formal modes of hearing – towards a wider range of situated, complexity-sensitive epistemic fora – but that traditional legal process values themselves might be re-imagined. Thus *audi alteram partem* and *nemo judex in sua causa* could be reconstituted:

> [t]he central point is that legal processes can become more sensitive to the multiple implications of the materio-semiotic knottings of Haraway's rich invitation to 'meet', as it were, between species of all kinds ... Science could inform this process by directing a whole range of non-human 'inputs' into ... an open juridical epistemology, producing a more eco-inclusive notion of *audi alteram partem* responsive to measurements of ecological integrity,[1] but resisting assumptions concerning the completeness or epistemic dominance of scientific knowledge itself. Such inputs might produce a reconceived '*alteram*' – the '*alter*' (or 'other') no longer being merely confined to a human or other existing legal protagonist (a company, a local authority and so forth), but being understood to embrace multiple life-forms and systems seen as *co-participants* in the formation of normative outcomes.

Grear's conclusion is apt as a conclusion for this collection:

> A new juridical imaginary is possible – and vital to build. The ... foundations are being laid ... Law and legal theory, by responding to ... new, passionate visions of an interwoven world, would follow ... a deeply ethical impulse. The time has come. New future histories summon the legal ... imagination – and potential new interpretations and modes of legal [and other modes of response] offer meaningful, practical ways of responding to the call.

---

[1] L Westra, K Bosselmann and R Westra, *Reconciling Human Existence with Ecological Integrity: Science, Ethics, Economics and Law* (Earthscan 2008).

# PART 1

# Philosophical investigations

# 1. Doubt and denial: epistemic responsibility meets climate change scepticism*

**Lorraine Code**

## 1. INTRODUCTION

Three themes run through the questions I will address in this chapter: my purpose is to suggest some insights that emerge from exploring their interconnections. First, I will read a stand-off between ecological thinkers/activists, climatologists, and climate-change sceptics as one way of indicating how questions central to the epistemology and politics of testimony bear, directly and indirectly, on debates surrounding climate crisis, gender justice, and development projects. In so doing, I use the label 'ecological thinkers' to refer to theorists and activists who work from an ecologically-informed stance towards understanding climate

---

* Originally published as L Code, 'Doubt and Denial: Epistemic Responsibility Meets Climate Change Scepticism' (2013) 3/5 *Oñati Socio-Legal Series* 838–853. Available from: http://ssrn.com/abstract=2247830. Lorraine Code is Distinguished Research Professor Emerita in Philosophy at York University in Toronto Canada, and a Fellow of the Royal Society of Canada. She is the author of *Epistemic Responsibility* (Hanover NH, University Press of New England 1987); *What Can She Know? Feminist Theory and the Construction of Knowledge* (Cornell, Cornell University Press 1991), and *Rhetorical Spaces: Essays on (Gendered) Locations* (London, Routledge 1995). In *Ecological Thinking: The Politics of Epistemic Location* (New York, Oxford University Press 2006) she develops an 'ecological naturalism', which draws on ecological science where Quineans look to cognitive science, as a place – literal and metaphorical – where knowledge is 'naturally' made. She is currently working on issues of gendered subjectivity as they bear on testimony, ignorance, and vulnerability, with particular reference to climate change scepticism. This research is funded by a grant from the Social Science and Humanities Research Council of Canada. The project's title is '"Manufactured Uncertainty" and Epistemic Responsibility: Implications for Climate-Change Skepticism'.

change, where thinking ecologically encompasses broad considerations of the implications and effects of historical-geographical 'situatedness' for citizenship and politics, broadly conceived. It is about critically imagining, crafting, and endeavouring to enact principles of ideal cohabitation with one another, and in and with the wider world. In taking my point of entry from a contrast between climate change sceptics and ecological activists, I am not presuming that this contrast alone is definitive of the matters I will address, but that framing the questions thus facilitates one plausible way of approaching the specificities and generalities of these multiply tangled issues.

In the stand-off I refer to, the debate often comes down to a contest – to fighting science with science – and, for lay persons, to a contest between/among conflicting expert testimonies in which decisions about where it is reasonable to place belief and trust are at once epistemologically and ethically-politically fraught. Secondly, then, I will propose that evaluations of testimony prompt vital questions about subjectivity and agency, and in this case also about freedom. Yet the climate-change-sceptical positions I will address presuppose an abstract, impersonal conception of autonomous gender-neutral subjectivity, which despite the neutrality of its generic presentation, is inevitably raced, gendered, and otherwise socially-politically-economically inflected and situated in ways that, I will suggest, are philosophically significant. The implications of their situatedness require closer scrutiny. Thirdly, to illustrate one way in which these lines of thought come together, I will briefly revisit Rachel Carson's ongoing *ad feminam* treatment by the science sceptics of her time, and still in ours. Her 'case', and the contest of testimonies it continues to generate, exposes some of the sexed-gendered specificities concealed beneath the façade of the presumptively generic epistemic and moral subject.

## 2. 'MANUFACTURED UNCERTAINTY' – MERCHANTS OF DOUBT

Climate change sceptics stand in a starkly oppositional relationship to ecological thinking,[1] to larger commitments to environmental sustainability, and to assumptions about subjectivity and the politics (and ethics) of

[1] L Code, *Ecological Thinking: The Politics of Epistemic Location* (Oxford, Oxford University Press 2006).

knowledge that inform these modes of thought. Their scepticism trades on a certain cynicism about knowledge, the earth/the world and people's lives, in the name of a profit-driven but also curiously conceived *freedom*-promoting-and-preserving instrumentalism. I intend 'cynicism', to capture a generalised, if tacit, distrust of professed ethical and social values in mass society, especially in circumstances where high expectations of institutions and human authorities seem often to go unfulfilled. It can manifest in a generalised disillusionment and distrust of putatively ameliorative social movements, organisations, and authorities; and it is commonly animated by and animates a starkly atomistic, self-contained and detached conception of subjectivity in practices and rhetoric that oppose ecologically informed social-political regulatory measures to evaluate and limit climate change, from a conviction that they are designed to tamper with, and indeed jeopardise, individual human freedom, itself cast as an incontestable entitlement and goal.

In what follows, I take as my point of departure a single text: Naomi Oreskes and Erik M Conway's 2010 *Merchants of Doubt*, which is pivotal to this discussion for its analysis of scientific questions in ways that speak to specialists and to an informed 'general public' alike, and for how its analysis is situated within an extensive yet subtle critique of capitalist free market society.

Concentrating on this book makes sense for the position I will take because of its unusual approach, through the history of science, ethics, politics, while foregrounding American exceptionalism. For epistemology and the politics of knowledge, the book is startling in its exposure of the curiously delusionary insistence, by certain deniers among US scientists and politically invested others, on assigning equal weight to 'both sides' of climate change debates, before approving regulatory measures. Yet Oreskes and Conway show that the 'sides' do not have equal claim to credibility, epistemic responsibility (the dominant issue here), or scientific integrity. Hence there is a need, at once, to understand and address the science, and to address the weight of an ideological opposition masked as a commitment to fairness and equality – dressed up in a uniquely American fear of governmental 'interference' with a quintessential US value: Liberty.

Sparked by Frederick Seitz's contention that global warming was simply a scare tactic, the result of pandering to irrational fears of environmental calamity by scientists seeking fame and fortune, the authors document a quintessentially neutral cognitive stance that trades on a conception of 'science' – homogenised and reified. This conception dispassionately sustains a level of doubt, and indeed ignorance, in a dominant, predominately white western society and population. For this populace the claim 'science has shown' carries a cachet enhanced by such presumed neutrality, which suggests that the facts alone are so compelling that reason cannot withhold assent. The veil of neutrality behind which this impersonal claim is articulated screens out any idea that specifically situated, and not disinterested, human subjects have produced the 'science' invoked, or that the effects of what 'science has shown' will be enacted in specifically situated, thence not disinterested or homogeneously implicated lives. Such inquiry works from a taken-for-granted assumption of human sameness. Yet no articulated conception of epistemic or moral-political subjectivity informs the inquiry; and the epistemic practices the authors depict are coldly inhospitable to examining how subjectivities are differentially constituted and enacted in this bleak picture of a social-ecological order where it is hard to see how anyone could be 'at home'.

That climate change deniers are motivated by financial gain will come as no surprise. More striking, philosophically, are their fears about the *loss of freedom* ecologically-informed regulatory measures will presumably entail. According to the Oreskes and Conway account, such fears tend to override financial concerns in significance, in twenty-first-century USA and indeed throughout the affluent populations of the western/northern world. The going concern (albeit with regional and demographic variations) is that if threats to 'the environment' require regulatory measures, then they interfere intolerably with human freedom – with Liberty – and should not be contemplated. The conception of freedom operative here is at the core of the autonomous self-contained and self-sustaining individualism that infuses the liberal democracies of the white capitalist western/northern world, whose masculinist-patriarchal enactments and effects – its demeanour of neutrality notwithstanding – have long been the subject of feminist critique. In a free society, the argument goes, 'we' will – *ex hypothesi* – refuse to be constrained in our freedom to smoke, pollute, or to use pesticides when and as we will. This refusal attests to a governing conception of subjectivity according to which, generically and neutrally, there can be no contestation of the freedom at issue, no question about its attainability, its everyday observance, and its viability as a sacrosanct human value. There is no interest in who these perfectly free subjects are, or in what ways sexed, racial, class-specific or any other differences from a white sex-gender-neutral norm could have played a constitutive part in shaping the communities, environments, and interactions where such freedom is defended. This conception of freedom is sharply individualistic – the freedom of the 'buffered self'[2] of masculinist modernity – born of the imagined self-sufficiency integral to a frontier mentality. It is nourished not by an imaginary of mutuality and responsible dwelling, but of entitlement, acquisition, and instant gratification. Such freedom is unsustainable in itself, especially in looming crises of scarcity, and it is a freedom without the resources to sustain even the subject who adamantly affirms it as his.

According to Oreskes and Conway, a conviction has prevailed in post-1970s America that there could be 'no freedom without capitalism and no capitalism without freedom'.[3] When it began to emerge that industrial emissions were damaging human and ecosystem health and that 'regulating' such emissions – which, allegedly, 'flew in the face of the capitalist ideal'[4] – could be the only responsible response, the struggle turned into a battle of science against science. Thus, for example, against the (US) Environmental Protection Agency's efforts to expose the harms produced by tobacco smoke, the argument was made that such proposals merely allowed 'bad science to become a poor excuse for enacting new laws and jeopardizing individual liberties' – 'It wasn't just money at stake, it was individual *liberty*' – 'Today, smoking, tomorrow… who knew? By protecting smoking, we protected freedom'.[5] In the controversy over multiple forms of pollution, the argument was advanced that if science was working against 'the blessings of liberty' 'they would fight it as they would fight any enemy. For indeed', Oreskes and Conway observe, 'science *was* showing that certain liberties are not sustainable – like the liberty to pollute'.[6] It is in this climate of resistance to knowing/acknowledging immanent threats of climate disaster that the politics of testimony I will discuss are being enacted, in what amounts to a struggle for ascendancy between an ecological imaginary and what Val Plumwood in *Feminism and the Mastery of Nature* aptly characterises as an epistemology of mastery.

---

[2] G Lloyd, *Providence Lost* (Cambridge, MA, Harvard University Press 2008) at 322.

[3] N Oreskes and EM Conway, *Merchants of Doubt: How a Handful of Scientists Obscured the Truth on Issues from Tobacco Smoke to Global Warming* (New York, Bloomsbury Press 2010) at 64–5.

[4] Ibid, 65.

[5] Ibid, 145.

[6] Ibid, 239.

## 3. ECOLOGICAL THINKING

A systemic prejudice against disrupting the complacency of the status quo is embedded in charges that climate change scientists are promulgating irrational fears. Such scepticism is a powerful force, perpetuated and insulated against condemnation by those for whom there are no immediate rewards for critically re-examining settled convictions that science-as-usual has all the answers, or for critically re-evaluating their everyday practices. For members of a social-economic elite, acknowledging and acting to minimise the injustices climate change scepticism condones would entail significant personal 'losses', not only of physical comfort, but of the myriad privileges and self-certainties that structure their entire ways of being. They need not take measures to reduce consumption, or to consume 'more sustainably' or more carefully, for the world has always 'healed' itself, and is bound to go on doing so. Why would people whose lives are constructed around the illusions such unsustainable practices uphold be prepared to relinquish these privileges that have long been theirs? This is the urgent question: it is at once epistemological, ethical, and political. Charges to the effect that climate change scientists are promoting 'irrational fears' highlight the intensity of the struggle between people who think and act within a hegemonic imaginary of mastery and entitlement and those committed to interrogating and unsettling it. The struggle is portrayed as being about fighting science with science, where 'good science' and 'bad science' tend to be categorised less according to the quality of the knowledge they produce, and more according to the interests they serve. In such circumstances, questions about epistemic responsibility rarely arise, and, despite their being germane to these inquiries, it is not easy to introduce them into the on-going deliberations.[7]

Consider one pertinent example: into a 1990s debate over risks posed by Environmental Tobacco Smoke (ETS) in North America, a publication *Bad Science: A Resource Book* was inserted.[8] Oreskes and Conway

---

[7] L Code, *Epistemic Responsibility* (Hanover NH, University Press of New England 1987).

[8] *Bad Science: A Resource Book*, BN: 2074144197, Legacy Tobacco Documents Library. Oreskes and Conway (n 3) at 147 observe: 'The phrases "excessive regulation", "over-regulation", and "unnecessary regulation" were liberally sprinkled throughout the book. Many of the quotable quotes came from the Competitive Enterprise Institute (CEI), a think tank promoting "free enterprise and limited government" and dedicated to the conviction that the "best solutions come from people making their own choices in a free marketplace, rather than government intervention"'.

describe it as 'a how-to handbook for fact fighters'.[9] Its intent was to challenge the authority and integrity of the US Environmental Protection Agency's (the EPA's) research into the effects of second-hand smoke. The contest is thus represented to the public as a debate between adherents to two equivalent, commensurable lines of scientific inquiry, where an awkward background presupposition prevails that 'science', curiously aggregated, is nonetheless a source of certainty. Nonetheless, each 'side' is driven by a specific agenda, neither of which most members, even of a well-educated lay public, are well qualified to judge. Yet, in western societies, non-scientists have little choice but to look to and endeavour to evaluate testimonial findings such as these to inform and animate their/our beliefs and conduct. The power of the 'narrators' is undeniable.

As Oreskes and Conway read them, the articles in the resource book, taken together, create an 'impression of science rife with exaggeration, mismanagement, and fraud'.[10] Hence, it is no wonder that the judging process for non-scientists is confusingly complicated, reliant as it is on multiple layers of conflicting testimony on the way to determining whom or what to trust. At best, in the conflict Oreskes and Conway detail, an ecologically committed lay reader may be able to discern a dispassionate if not disinterested approach in the EPA scientists' research, whereas *Bad Science* seems, at least in the reading the authors present, to rely on cautionary quips and throwaway lines about betrayals of public trust and costly policy decisions, infused before the fact with the very political beliefs the deniers are seeking to establish. (One 'MESSAGE' reads: '*Proposals that seek to improve indoor air quality by singling out tobacco smoke only enable bad science to become a poor excuse for enacting new laws and jeopardizing individual liberties*').[11] Such impressions scarcely suffice as a basis for informed judgement. Oreskes and Conway acknowledge that *Bad Science* does quote 'experts', noting however that many of these experts were paid consultants to industry: an involvement that is rarely, and then often obliquely, announced in their assessments. In the main, though, the text adopts 'a more sophisticated strategy: reminding readers of the fallibility of science'.[12] A similar strategy is evident in some of the essays in the aptly-named text

[9] Oreskes and Conway (n 3), 6.
[10] Ibid, 146.
[11] Ibid, 144–5, italics in original.
[12] Ibid, 146.

*Agnotology: The Making and Unmaking of Ignorance*[13] whose purpose, with variations from essay to essay, is to expose and unmask corporate and other attempts to promote and sustain ignorance of the effects of anti-ecological thought and action. Such attempts trade on that margin of fallibility, of uncertainty, to promote a level of doubt that would caution against initiating or endorsing regulatory measures. The tobacco industry, pharmaceutical companies, and climate change deniers figure prominently in these analyses. Thus for example, in the chapter from which I borrow the title of this chapter, David Michaels contends '[u]ncertainty is manufactured. Its purpose is always the same: shielding corporate interests from the inconvenience and economic consequences of public health protections'.[14] Either way, and this is the crucial point, the judicious response cannot plausibly be to conclude that because science is frequently marked by a degree of uncertainty, is indeed fallible, there are no good reasons to heed its warnings. Further inquiry, deliberation, debate has to remain on the agenda.

For non-scientists and members of the larger population who often, despite their best efforts, cannot claim a level of scientific literacy sufficient to enable them to evaluate the contradictions between climate sceptical and 'science as usual' positions, the issues are frustratingly puzzling. One possible approach in addition to on-going individual and collective investigation, I am suggesting, is indeed to initiate public conversations, consultations, and collaborative debates towards developing better informed *advocacy* processes which could facilitate navigating a world that is increasingly incomprehensible to non-specialists, non-experts. Such a proposal will meet with resistance from those for whom advocacy is by definition a tainted practice: Michaels, for example, asserts: 'Opinions submitted to regulatory agencies by corporate scientists ... must be taken as advocacy, primarily, not as science'.[15] Yet this comment presupposes a false dichotomy between science as knowledge-conveying and advocacy as mere propaganda, where surely the problem under consideration is about advocating responsibly or otherwise, and where *epistemic* responsibility – the responsibility to know well in order to advocate honourably – must figure largely. Learning where and how, reasonably, to trust testimonial evidence that one may understand only

---

[13] RN Proctor and L Schiebinger (eds), *Agnotology: The Making and Unmaking of Ignorance* (Stanford, Stanford University Press 2008).

[14] D Michaels, 'Manufactured Uncertainty: Contested Science and the Protection of the Public's Health and Environment' in Proctor and Schiebinger (n 13) 90–107 at 96.

[15] Ibid, 102.

partially is the hardest task: advocacy can contribute well or badly to the project of doing so, but it cannot be presumed villainous or virtuous before the fact. Clearly, at the very least, communal, collaborative deliberation has to be an open option here, for epistemic individualism comes up against its limitations in such endeavours.

Against the backdrop of a tacit yet widespread conviction that the epistemological questions in these debates are generically *human*, to which neither sex, nor race, nor any other specificities of 'identity' pertain, I turn to my second concern: about subjectivity and agency, which I will propose are constitutive, if often silently so, of practices of giving and receiving testimony. Sex-gender issues may seem to be quite irrelevant to these debates: the *Merchants of Doubt* text tacitly presents them thus; it is not unusual in so doing, and its political motivations are nonetheless laudable for ecological thinking. Still, I am contending that there are valuable, activism-informing insights to be gained from taking subjectivity into account in these matters, especially given the tenacity of a residual, unreconstructed public image of 'science' as impersonally objective. The 'subject' – hence the subject seeking to defend *his* freedom – makes few explicit appearances in the *Merchants of Doubt* text, even though the scientists on both 'sides' are mentioned by name and credentialed accordingly. Yet the recurrent worries about freedom that run through this story expose the effects of a tacit conception of subjectivity that merits closer investigation. It is too rarely observed that at least in white western societies, the freedoms abstractly invoked and zealously defended by the climate change deniers are not equally distributed across the sex/gender – or any other – social-political order. (Recall Val Plumwood's reminder that women '… provide the environment and conditions against which male "achievement" takes place, but what they do is not itself counted as achievement'.[16] In these debates, tacit conceptions of 'freedom' and 'achievement' are so closely aligned as to be co-constitutive.) The point, too tersely put, is that expectations of and assumptions about freedom and constraint take for granted and are informed by certain enactments of subjectivity that are neither neutral, nor universally realisable, nor politically innocent. Even when such enactments of subjectivity remain tacit, they are integral to producing and sustaining social-epistemic practices to such an extent that they cannot be left unaddressed.

---

[16] V Plumwood, *Feminism and the Mastery of Natur*e (London, Routledge 1993) at 22.

Feminists and other marginalised Others from an unmarked white affluent masculine norm have long insisted that all freedom is someone's freedom, just as all knowing is someone's knowing, where someone can be singular or plural – or singular because plural – but either way, it matters for the production, circulation, uptake, and evaluation of knowledge claims. Feminist epistemologists have been engaged since the early 1980s in showing, to varying degrees, how the 'sex of the knower' plays a constitutive part in these processes:[17] it is within the space their/our analyses open that I am reading Rachel Carson's life and work. Evidently, the freedom the merchants of doubt are determined to protect is principally available, in affluent western societies, to propertied heterosexual not-yet-old white men. Hence, 'taking subjectivity into account'[18] in thinking about how ecological matters are known requires taking gendered/raced/classed subjectivities into account critically and genealogically, in the specificities of their time and place. It requires examining how the doing, knowing, being, dwelling at issue are enacted there, and analysing the implications of 'situatedness' for knowledge projects. As Plumwood also reminds us, integral to the social imaginary that sustains mainstream Anglo-American philosophy is an assumption that 'the human' is implicitly masculine, not just conceptually but in its effects, 'while the feminine is seen as a derivation from it'.[19] Likewise, the contestable ideal of 'autonomy' underlying the pleas for freedom Oreskes and Conway report has, throughout the history of western liberal political thought, derived from and celebrated possibilities afforded by and to affluent white male/masculine lives.[20] Even after decades of feminist theory and practice these considerations remain urgently under-acknowledged. I am proposing that they need to be critically revisited, especially when appeals to a generic *freedom* are invoked to excuse or

[17] Space does not permit an elaboration of these claims, but feminist epistemologists have, since the 1980s, advanced sophisticated, nuanced arguments in support of this assertion. For state-of-the-art analyses, see the essays in H Grasswick (ed.), *Feminist Epistemology and Philosophy of Science: Power in Knowledge* (Dordrecht, Springer 2011). This reference to 'the sex of the knower' is to L Code, 'Is the Sex of the Knower Epistemologically Significant?' (1981) 12 (3 & 4) *Metaphilosophy* 267–276.

[18] L Code, 'Taking Subjectivity into Account' in L Alcoff, and E Potter (eds), *Feminist Epistemologies* (New York, Routledge 1993) 15–48.

[19] Plumwood (n 16) at 23.

[20] For a pertinent critique of autonomy see L Code, 'The Perversion of Autonomy and the Subjection of Women: Discourses of Social Advocacy at Century's End' in C Mackenzie and N Stoljar (eds), *Relational Autonomy* (New York, Oxford University Press, 2000) 181–211.

condone ecologically destructive practices. As to how this revisiting would evolve and what its effects might be, it will in large part be a deconstructive and genealogical project of exposing the sources and power-infused social-political effects of tacit yet entrenched assumptions about whose knowledge matters and can claim acknowledgement, what kinds of knowledge achieve hegemonic status, and which knowledge claims rightly count as exemplary for epistemology and beyond. Such questions are posed in several articles in the *Agnotology* volume.[21] They may not seem to matter for such self-contained, detached examples as the cups on tables or cats on mats beloved of classical empiricist inquiry, or for the fake barn facades of the new empiricism, but they do matter for social knowledge that informs or thwarts ecological thinking and practice, in ways that reinforce entrenched injustices; and they do matter for projects of critical engagement with the politics and epistemology of testimony.

From an ecologically-informed stance Mick Smith, in *An Ethics of Place*, sums up the troubling situation I have been discussing. I quote him at length:

> As modernity's offspring we … tend to understand our own identities and social relations … [through an] atomistic ideal/ideology of autonomous, bounded, individuality … [as] concrete and isolable individuals each on their own disparate trajectories, each with *particular* identities derived from … certain essential, quantifiable, and indefeasible properties. We are born under the sign(s) of one-dimensional *man*; *Homo economicus*, that self-contained and self-serving caricature of modern humanity, a parodic recapitulation of the instrumental order of capitalism and phallocracy. … Only 'man' is intrinsically valuable. Women and nature are made subject to reason's cold calculations, their reality recognized only insofar as they become hard currency to be valued and traded according to their use.[22]

Taking such thoughts seriously for their ethical-political and epistemological import, as I am proposing we do, calls for critically renewed conceptions and enactments of subjectivity at a social, collective

[21] Proctor and Schiebinger (n 13). Relevant in this regard are the articles by Charles Mills, Londa Schiebinger, Nancy Tuana, and Alison Wylie in that publication.

[22] M Smith, *An Ethics of Place: Radical Ecology, Postmodernity, and Social Theory* (Albany, State University of New York Press 2001) at 171, italics in original).

level: enactments distantly yet pertinently analogous to late-twentieth-century consciousness-raising practices. They require rethinking/re-enacting who *we* are, in ways sufficiently discerning to unsettle such deeply sedimented convictions. This is the hardest requirement: it is easier, more imaginable, to think and engage in revisionary ways of doing, thinking, knowing. But to practise a philosophy which requires rethinking and re-enacting basic assumptions about *who we are* is ontologically and epistemologically radical, upheld as these assumptions are by an instituted social-political-epistemic imaginary in which as inhabitants of the affluent west we (often unconsciously) live and think, however obliquely or contrarily. Nor is the issue quite as Smith puts it in urging an '*alternative* conception of subjectivity',[23] for the language of 'alternatives' is misleading in its implication that all ways are up for grabs: *we* can opt for one or another, interchangeably and intermittently, as we would select from a smorgasbord of edibles. An activism that follows from Smith's recommendations would work singly and collectively to unmask, discredit, and displace that 'caricature of modern humanity', exposing it for the dangerous illusion it is through on-going, piece-by-piece deconstructions of its contributions to producing the ecologically unsustainable conditions that prevail in affluent parts of the western/northern world: to undermine its ontological and – in consequence – its epistemic credibility. In my view, feminist theory and practice, both singly and in their alliances with other post-colonial movements of the late twentieth and early twenty-first centuries, are engaged in just such displacement projects, and at their best are indeed aware that they have to contest who *we* are as fundamentally as they contest the social-economic structures we make and are made by. They/we have not yet completed this project, but they/we cannot desist.

Finally, as I have suggested at the beginning of this chapter, the example of Rachel Carson is instructive for examining connections between expertise, testimony, and subjectivity. In *Ecological Thinking*, I introduce her as an exemplar of *ecological subjectivity* whose way of inhabiting the world confers content on the idea and the practice in ways consonant, among others, with Verena Conley's claims for ecological subjectivity as 'relating consciousness of the self to that of being attached to and separated from the world'.[24] For Conley the task for ecological thinking is to unmask 'mass-produced subjectivity in societies of control

[23] Ibid, at 173, my italics.

[24] VA Conley, *Ecopolitics: The Environment in Poststructuralist Thought* (London, Routledge 1997) at 10.

with their consequences for natural and social ecology'.[25] As I enlist them here, these thoughts also contest unexamined ontological presumptions of human sameness, freedom, and the autonomous liberal unified self: in so doing they are as normative as they are descriptive. Yet it should also be borne in mind that even though the natural world, to a considerable extent, constrains the approaches and points of view that can achieve consensus within a discipline, in interpreting and understanding ambiguous data and managing uncertainty, as Eric Biber rightly notes: '[w]here reasonably possible, scientists tend to interpret their observations as consistent with whatever theory currently commands the most adherents, even if other interpretations are equally or even more plausible'.[26] As theories of ecological subjectivity recognise, such perspectives are as often influenced by the values of the members of the discipline as by situational factors. Those values, too, have to be held open to analysis and critique, but the fact of the influence is not in itself reprehensible.

Carson is not just the ecological subject but, in ways pertinent to the claims I am advancing here, she is also (to borrow Luce Irigaray's term) the *sexuate* subject, albeit, in the climate of her times, often silently, tacitly, and with positive and negative implications.[27] It is clear that Carson herself regarded neither her way of life nor her scientific practice as shaped or indeed otherwise influenced by her female/feminine being. Indeed, no conceptual-discursive space was available for addressing such issues, either during her lifetime, or for some years after her death.[28] Yet within the discourse of feminist and post-colonial epistemology and

---

[25] Ibid, 74–75.

[26] E Biber, 'Which Science? Whose Science? How Scientific Disciplines Can Shape Environmental Law' (2012) 79(2) *The University of Chicago Law Review* 471–552 at 503.

[27] L Irigaray, *Philosophy in the Feminine* (London, Routledge 1991) at 50. Consider Margaret Whitford: '… to speak or write like a man is to assert mastery, to claim truth, objectivity, knowledge, whereas to speak like a woman is to refuse mastery, to allow meaning to be elusive or shifting, not to be in control, or in possession of truth or knowledge. So to be assertive, to make claims, to be "dogmatic", which means to have a thesis, a meaning, a political position, is to take up a "male" stance, whatever one's sex': M Whitford, *Luce Irigaray: Philosophy in the Feminine* (London, Routledge 1991) at 50.

[28] The allusion is to Michel Foucault's observation: 'Mendel spoke the truth, but he was not *dans le vrai* (within the true) of contemporary biological discourse …': M Foucault, 'The Discourse on Language', *Appendix to The Archaeology of Knowledge* (Translator AM Sheridan Smith) (New York, Pantheon 1972) at 224.

philosophy of science now, her scientific practice, for all the regularity of its approach and competence, hovers on the edge of attesting to a feminist standpoint, and is often discredited accordingly.[29] Some of her best, most ecologically sophisticated work sits just here, in a not-yet-realised sex-gender-specific frame and style of reasoning.[30] It is she who exemplifies the power and the perils of an informed advocacy that, I have insisted, is a crucial piece of sound ecological practice. (Her style contrasts sharply with the *Bad Science: A Resource Book* agenda.)

For a time, Carson achieves impressive scientific and public legitimacy and acclaim, if only uneasily and precariously. She dies from breast cancer: from one of the ecologically most powerful (putative) effects of the substances she studies.[31] She is the only epistemic subject who claims a fully narrated place in the *Merchants of Doubt* text: one who is (ambivalently) celebrated in her time yet is after the fact vilified: castigated as a murderer.[32] In ways I will explain, she is again even now being cast out as the sacrificial subject. That said, it must be noted that Carson is not the only scientist identified by name in the Oreskes and Conway text, nor the only one whose qualifications and training are cited

---

29 See A Wylie, 'Why Standpoint Matters' in R Figueroa and S Harding (eds), *Science and Other Cultures: Issues in Philosophies of Science and Technology* (New York, Routledge 2003) 26–48. See also H Doremus, 'Listing Decisions under the Endangered Species Act: Why Better Science Isn't Always Better Policy' (1997) 75(3) *Washington University Law Quarterly* 1029–1153. Available from: http://openscholarship.wustl.edu/cgi/viewcontent.cgi?article=1579&context=law_lawreview.

30 'Styles of reasoning' is Ian Hacking's phrase: I Hacking, 'Language, Truth and Reason' in M Hollis and S Lukes (eds), *Rationality and Relativism* (Cambridge MA, MIT Press, 1982) 48–66.

31 LH Sideris, 'The Ecological Body: Rachel Carson, Silent Spring, and Breast Cancer' in LH Sideris and KD Moore (eds), *Rachel Carson: Legacy and Challenge* (Albany, State University of New York Press 2008) 136–148.

32 For example, Oreskes and Conway note that (as of 2007) '[t]he Internet is flooded with the assertion that Carson was a mass murderer, worse than Hitler. Carson killed more people than the Nazis. She had blood on her hands, posthumously. Why? Because *Silent Spring* led to the banning of DDT, without which millions of Africans died of malaria' (Oreskes and Conway (n 3) at 216). Noting that 'her legacy has been characterized as "Rachel Carson's Ecological Genocide"', Maguire cites the relevant article: '… let there be no mistake: Rachel Carson and the worldwide environmentalist movement are responsible for perpetuating an ecological genocide that has claimed the lives of millions of young, poor, striving African men, women, and children, killed by preventable diseases': S Maguire, 'Contested Icons: Rachel Carson and DDT' in Sideris and Moore (n 31) 194–214 at 194.

to validate their testimony: the climate change defenders and deniers, too, are identified with their credentials. Yet their place-holder status in claiming 'truth, objectivity, knowledge' is neutrally assumed, despite bitter disagreements about their findings; and they appeal to or deplore threats to a freedom that is presumptively everyone's and no one's, individualistically and impersonally defended.

What then is my purpose in singling Rachel Carson out for discussion? Although, as I have said, most of the scientists whose work is discussed in the *Merchants of Doubt* text are indeed named and their credentials and allegiances detailed (even some of those they try to conceal), Carson is the only one who *in herself* – as contrasted with *for her work* – becomes a focus of attention. In the politics of knowledge, this contrast is significant. It can scarcely be a coincidence that the only female environmental scientist discussed at length in the book is one who appears to have been judged as much by her 'person' as by her work, that analyses, both critical and otherwise, seem unable to avoid reference to her sex or to refrain, implicitly or explicitly, from factoring it into their evaluations, usually negatively, if often only obliquely. I emphasise this point not to charge Oreskes and Conway with having offered a sexist reading, but to note, if cautiously, that Carson herself is more plainly visible in her work than are many of the scientists they discuss, that her gender in this respect is pertinent, and that in the scientific circles where she was active it could as readily have been presented as the focus of laudatory as of condemnatory or silently dismissive discussion.

It need scarcely be said that there is a peculiar vulnerability to the situation of a woman in science, even more than a full decade into the twenty-first century, and dramatically more so for a woman of Carson's time. Such *personal* vulnerability, again, is well known and not difficult to imagine. Less well known and, I suggest, still more significant, is the *epistemological* vulnerability that attends it or grows out of it. Sex-specific charges against Carson's work and thought were usually covered over in her time, yet they were there, if sometimes veiled – as they are again now in critiques too numerous to recount. Accusations of hysteria proliferate. In the Oreskes and Conway chapter, where the charges are multiply detailed, such accusations have as much to do with Carson's sex as with epistemically irresponsible accusations that she was doing bad science – yet where, on closer examination, it becomes apparent that the irresponsibility has less to do with the substance of her research and more to do with the attackers' and discreditors' failures to base their charges on adequate investigations of the on-going evolving complexities of the effects, for example, of DDT – which are neither as uniform nor as static as the 'evolution' of rocks might claim to be. Still more germane to the dismal quality of the critics' assessment of her research than their having made a scapegoat of Carson herself is their failure to take into account such effects as 'the well-documented and easily found (but extremely inconvenient) fact that the most important reason that DDT failed to eliminate malaria was because insects *evolved* ... a truth that those with blind faith in free markets and blind trust in technology simply refuse to see'.[33] This rush to judgement is epistemologically reprehensible. Oreskes and Conway observe, for example, that to know whether the deleterious effects of DDT for female reproductive health were significant enough to support continuing to ban it, it was vital to engage in longitudinal studies, for example, of women who had been exposed early in life, 'when environmental exposures where high'.[34] One such study, conducted in 2000–2001 on women then in their 50s and 60s who had been exposed to DDT as children or teenagers 'showed a fivefold increase in breast cancer risk among women with high levels of serum DDT or its metabolites'.[35] If such findings can translate across the weakened effects of DDT in malarial districts now, then they contribute to vindicating Carson's longitudinal approach, even as they urge a further departure from the instantaneous spectator epistemology of standard empiricism towards 'horizontal' as contrasted with vertical, top-down analysis: to taking a longer view across terrains and time-frames, when the subject-matter is appropriate.

By contrast, characteristic of Carson's scientific 'style' is her wariness of too-ready a translation from one domain or species to another. Although in her scientific practice she undoubtedly understands the allure of mastery and control, her ways of achieving it often require painstakingly following up narrow and precise local hypotheses that differ significantly for each of the species she studies – as is apparent in her research on the Japanese beetle, the gypsy moth, Dutch elm disease; as the hypotheses differ also for investigating the short- or long-term implications of chemicals for human health. Catching a central contrast between an overarching ethos of mastery and an ecological ethos, Carson deplores a stubborn corporate and more widespread resistance to taking a longer view – to waiting 'an extra season or two'[36] – when a quick (chemical) fix is ready to hand. Ecologically, thinking as she does requires factoring time, place, and history into responsible scientific investigations. As I understand it, this is a large, not a small

33 Oreskes and Conway (n 3) at 236, italics original.

34 Ibid, 229.

35 Ibid.

36 Code (n 1) at 42.

epistemological claim: it bears directly on her thinking about causality, where she also departs from standard epistemological assumptions, and challenges them in so doing. And it points towards the need for closer investigations of issues of *time* in the philosophy of science, ecological thought, and climate change epistemology. This is a delicate point to endorse unequivocally in climate change discussions, for despite the claim's initial plausibility, climate change supporters insist, and rightly, upon the urgency of immediate action. They argue that whatever the uncertainty, action is required, now, based on the best available predictions. 'We' may not have even a season to wait. Carson, in the 1950s and 60s was working from the (perhaps tacit) conviction that causal connections may not be immediately obvious to people neither informed – nor prepared – to look ecologically. Thus it is easy to discount causal claims that extend temporally and geographically away from a chemical application; and that require imagination, conjecture, patience and time for confirmation or falsification. Her ecological approach vindicates just such a longer view. Yet the hypotheses she works from can, in some situations, guide inquiry whose empirical generalisations stand up well against the quick and dirty solutions proposed by the chemical industry and its champions.

Whether Carson's diverse achievements can be attributed to her being a woman is a different and more complex question. Many aspects of her scientific practice and her writing style, some of which I have mentioned, contrast sharply with the orthodoxy of received 'scientific' practice. But attributing gendered significance to them is not a straightforward matter, even though her work displays affinities with explicitly feminist epistemological inquiry. More to the point, however, and again germane to some of the after-the-fact gendered claims, is the extent to which Carson was often criticised for her *styles* of research and reasoning rather than for their substance, and the extent to which the negative consequences of the substance of her research were laid at her feet as explicitly *ad feminam* charges. Carson, as Oreskes and Conway note, 'documented at great length both the *anecdotal* and systematic scientific evidence that DDT and other pesticides were doing great harm'.[37] Ironically, her respect for the 'anecdotal', which is also integral to her practice, has contributed to the fluctuating respect and easy vilification in the condemnatory rhetoric of her detractors. It connects with the rhetorical, conceptual architecture of the epistemological-scientific world where Carson worked, structured as it was by hierarchical divisions between fact and

---

[37] Oreskes and Conway (n 3) at 219, my italics.

anecdote, truth and narrative, reason and feeling, of which the first item in each pair claims greater public and professional credibility, authority, and reliability than the second. The division locates anecdote, narrative, and feeling on the negative, subjective, feminised side where meaning can be 'elusive or shifting', while for epistemologies of mastery it is in *facts* alone, dispassionately discovered, that truth is to be found. In her respect for down-on-the-ground experiential reports commonly dismissed as merely *anecdotal*, Carson's epistemic practice unsettles these distinctions – and earns her the label 'subversive' in so doing.[38] The larger issues in thinking about testimony in conflict are about how credibility plays out in the politics of knowledge, how credibility has been tacitly coded 'masculine', and how thought styles and styles of reasoning tainted with feminine associations can claim a place within 'the credible', 'the rational' only by following its formal, disinterested dictates. Rereading Carson's work in 2015 and evaluating its struggles to achieve acknowledgement and uptake when the epistemology and politics of testimony are animating vibrant new developments in social epistemology, suggests that her reliance on what was dismissively labelled 'anecdote' might now more readily have assured her a 'respectable' place in new approaches to public knowledge and ecological investigation.[39]

Although in her time such discrediting as I have mentioned was less often conveyed in sexed/gendered terms than it subsequently has been, the flavour was unmistakable. Now, it is less carefully masked. So for example, Michael Smith, in an article tellingly titled 'Silence, Miss Carson' (a title he borrows from an 'unbalanced' review of *Silent Spring*) notes a 'prevailing attitude' for which she was 'an uninformed woman who was speaking of that which she knew not. Worse, she was speaking in a man's world, the inner sanctum of masculine science in which, like the sanctuary of a strict Calvinist sect, female silence was expected'.[40] Nor is it merely a coincidence that the most damning website devoted to discrediting Carson, which comes from a 2009 Competitive Enterprise

---

[38] MH Lytle, *The Gentle Subversive: Rachel Carson, Silent Spring, and the Rise of the Environmental Movement* (New York, Oxford University Press, 2007).

[39] In this regard, see L Code, 'Testimony, Advocacy, Ignorance: Thinking Ecologically about Social Knowledge' in A Millar, A Haddock, and D Pritchard (eds), *Social Epistemology* (Oxford, Oxford University Press 2010) 29–50; M Fricker, *Epistemic Injustice: Power and the Ethics of Knowing* (Oxford, Oxford University Press 2007).

[40] M Smith, '"Silence, Miss Carson": Science, Gender, and the Reception of Silent Spring' in Sideris and Moore (n 31) at 172.

Institute project, is called rachelwaswrong.org.[41] Apart from the contents of the items on the site, the chastising 'bad little girl' tone of naming it thus is egregiously insulting: a woman too insignificant to be referred to by her full adult name has ventured too far onto territory that should not be hers, and is sternly reprimanded. This is the woman who in her time and still frequently in ours is dismissed as hysterical, or scare-mongering. In short, to understand Carson as 'a human being' and as a woman struggling to be a scientist in an inhospitable environment is to understand something of the *dis*-ease, the *un*-ease of that position: she is both an exemplary ecological subject – in part to her triumph – and an exemplar of a sexed-gendered being, if neither one always in her own eyes.

## 4. INTERIM CONCLUSION

In concluding I will again draw my three lines of thought together to confirm my reasons for pursuing them. First, as I understand it, the growing significance of testimony in social epistemology creates spaces for epistemologists and moral-political theorists to engage philosophically with situations in the division of intellectual-epistemic labour in western societies where 'we' are commonly reliant on other people to 'know' for us and where often, by this very fact, we do not and perhaps cannot know well enough to know whether they are knowing responsibly or to judge how to place our trust wisely. These thoughts engage with some of the most challenging issues in present-day epistemology. Thus, as I note at the beginning of this chapter, I have been reading the Oreskes and Conway text, together with analyses of Carson's scientific practice, as events in the epistemology of testimony.[42] The chains of analysis can be long and interwoven, and certainty may be elusive. But such is the consequence of breaking away from an individualistic picture of knowing towards a community or communal one, where neither the composition

---

[41] http://rachelwaswrong.org/. For a fuller account of Carson's scientific practice, see Code (n 1) especially chapter 2 and L Code, 'Feminism, Ecological Thinking, and the Legacy of Rachel Carson' in M Luxton, and MJ Mossman (eds), *Reconsidering Knowledge: Feminism and the Academy* (Winnipeg Manitoba, Fernwood Press 2012) 150–168.

[42] See, for representative texts, CAJ Coady, *Testimony: A Philosophical Study* (Oxford, Oxford University Press 1992) and J Lackey and E Sosa (eds), *The Epistemology of Testimony* (Oxford, The Clarendon Press 2006).

of the community nor our capacities to evaluate it well can be presupposed before the fact. Hence, secondly, I have urged acknowledging the need for critical analysis of some of the taken-for-granted assumptions about subjectivity that prevail in epistemic communities, and ascertaining the place of those communities in constructing and circulating public knowledge. Thirdly, when certain subjectivities stand out either as positive epistemic exemplars or as the reverse, it is instructive to examine, genealogically, how and why they claim or fail to claim that status, and by what warrant or withholding of respect. Thus, as I have noted, advocacy too has in many respects been a feminised practice; and it too meets with resistance whenever its combative, legalistic (masculinist) truth-denying reputation drowns out its emancipatory potential, as if people were too stupid to practice, judge and evaluate it well. People will argue for – advocate for – what they care about and/or fear to lose: both the deniers and the convinced engage in such practices, and advocacy good of its kind requires epistemically responsible, on-going investigation to inform it. It has to be kept open to justification or contestation at various levels. Yet the response need not, and I maintain *should not* be to condemn advocacy *tout court*, but to be vigilant for, engage critically with, and deplore, its blindly aggressive and/or wilfully ignorant instantiations. As educators we need to learn and teach how to advocate responsibly, knowledgeably, and humbly – paradoxical as this may seem – in the minutely informed and ethically/politically respectful way Oreskes and Conway investigate: to show how zealously the deniers seek to defend places and putative values that are simply unsustainable. Part of the bad press advocacy encounters has to do with an entrenched philosophical division between epistemology and ethics-politics that allows no space for the urgency of epistemic responsibility. Such a division, I suggest, is no longer plausible, now that social epistemology is demonstrating its effectiveness.

For the merchants of doubt, impersonal gender-neutral 'facts' are marshalled to promote or defend an impersonal, dislocated freedom. They can effectively be countered only in a complexly textured tale where, epistemologically, the task is at once phenomenological and hermeneutic. It is also empirical, but it is more effectively empirical when it is at the same time phenomenological and hermeneutic. There, the epistemological and social-political cannot be held apart. A doubled consciousness such as may have been Carson's but must also be ours will come insistently but differently to bear on each of the many subject matters, subjectivities, and issues that have to be investigated if sustainable futures are to be promoted against the insistence of merchants of doubt who are determined to gainsay them in the name of a freedom that is destructive at the core, even of the subjects who champion it.[43] Whether the denials can continue in the face of the 2013 IPCC report remains to be seen; the deniers will have to work relentlessly to manufacture levels of uncertainty sufficient to counter the force of these findings. Yet as Will Hutton observes: '[I]t will be met by a barrage of criticism from the new "sceptical" environmental movement … which, while conceding that global temperatures are rising, insists that there is still insufficient scientific proof to make alarmist predictions'.[44]

---

43 Some sections of this chapter are drawn with only minor changes, from my chapter entitled, '"Manufactured Uncertainty": Epistemologies of Mastery and the Ecological Imaginary' in P Rawes (ed.), *Relational Architectural Ecologies: Architecture, Nature and Subjectivity* (London, Routledge 2013), 73–90. They are reproduced here with permission of Peg Rawes and the Editors.

44 W Hutton, 'Our Planet Needs Us to Fight for its Survival', *The Guardian Weekly*, 189(16), 18.

# 2. Actors or spectators? Vulnerability and critical environmental law*

**Andreas Philippopoulos-Mihalopoulos**

## 1. VULNERABILITY

The old dialectics between 'us' and the environment that has nurtured the West and affected the rest of the world is not only outdated but positively dangerous. The polarisation has legitimised actions that have led to environmental degradation without, moreover, managing to explain how it is that the degradation affects human society. Since ecological disasters do not respect the conceptual limits of the polarisation, the latter had to be put on a more benign epistemological basis and reinvented along the lines of a new theoretical dialectics. Thus, instead of the rather grand ontological polarisation between 'us' and 'the environment', the debate has moved to an epistemological debate between anthropocentricity and ecocentricity. There are two problems with this move: first, this new theoretical language has provided the safety of a distance from the ontology of the human/non-human interaction, replacing ontology with an epistemology of supposed action; and, second, the new polarisation legitimised rather roundly, another perennial problem: that of 'the centre'. Indeed, while the ontological debate between 'us' and 'the environment' only insinuated the need for a centre (habitually assumed by humans), the epistemology of anthropo/ecocentricity plainly states that there is no alternative: however much we battle between extremes, we have no choice but to remain faithful to the soothing idea that there must be a centre. Why is this? It is largely because of fear. When Peter Sloterdijk[1] talks of the need of human societies for 'immunity', he

* This chapter was first published as A Philippopoulos-Mihalopoulos, 'Actors or Spectators: Vulnerability and Critical Environmental Law' (2013) 5/3 *Oñati Socio-Legal Series* 854–76.

[1] P Sloterdijk, *Sphären I: Blasen: Mikrosphärologie* (Frankfurt am Main, Suhrkamp 1998).

reveals the vast arsenal of concepts built around our fear of being exposed, of being thrown out into the open rather than hidden behind makeshift bubbles of immunity and spheres of fragile cohesion. Contrary to initial impressions, the centre is one such tool of avoiding exposure. While impressionistically in the limelight, the centre hides behind the blinding light and conceals its all-controlling position with a thick blanket of hinterlands, purely in the service of confirming centrality while avoiding exposure. There is no centre that is not surrounded by a floating periphery, capable of coating harsh edges in soft, invisibilising matter. One feels safe in a centre – the selective, controlled form of exposure that comes with it has the invigorating effect of full attention to 'us' as being the most important, most protected ones. Conceptualising the centre in this way, one quickly understands how the whole eco/anthropocentric debate builds on bravado rather than on the actual courage that a real exposure entails. This is the grand paradox of the centre: it looks exposed but it is actually the safest position to be in.

And there we have it: a continuous polarisation, fighting for a sound cause (if on the side of ecocentricity) but destined to fail in any transformative aim because of its reproduction of old dialectical, hierarchical semantics. Neither is this trap successfully resisted by environmental human rights discourses, since they are perhaps most acutely characterised by the tension between protecting the human centre while not divorcing this centre from an environmental consciousness. But the issue here is, once again, that all-too-human consciousness that trammels the 'surrounding' ecology in specific and, ultimately, anthropocentric ways. New semantics are needed that move away from the glib safety of the centre and these consequent, misaddressed dialectics. New semantics, I argue, must be sought on a surface of acentric continuum, traversed by velocities and pauses that do not easily subscribe to centralising generalisations. Following Gilles Deleuze and Félix Guattari[2] on this, I would call this surface 'the plane of immanence', namely the all-embracing sum of folds and falls and connections that contains its own origin, causality and teleology without transcendence. This plane operates on both an epistemological and an ontological level – it is both how we conceive of things and how things are. Just as it does not contain divisions between the observing spectator and the involved actor, this plane also offers no distinction along the lines of human/natural/artificial/technological. On the contrary, these categories fold into each other and constantly emerge

---

[2] G Deleuze and F Guattari, *A Thousand Plateaus: Capitalism and Schizophrenia* (Translation, B Massumi) (London, Continuum 2004).

as epistemological and ontological hybrids. It is in this sense that here, instead of actors, I would like to refer to actants, namely an entity that goes beyond the distinction human/non-human and consists, most often, of a combination of both. Bruno Latour defines them as 'something that acts or to which activity is granted by others. It implies no special motivation of human individual actors, nor of humans in general. An actant can literally be anything provided it is granted to be the source of an action'.[3] It is clear that on the surface where the distinction between human and non-human is blurred, 'actant' conveys the complexity of the human much better than the rather fraught 'actor'. In the same vein, the concept of 'spectator' implies a privileged position that enjoys the critical distance of the spectacle of the world. It is a peculiar position that humans have managed to construct for themselves, where the world performs for them. Guy Debord was writing at the beginning of last century: 'The society based on modern industry is not accidentally or superficially spectacular, it is *fundamentally spectaclist*'.[4] The society of spectacle constructs a full economy of spectacle where the environment is variably dressed up as resource (with all the consequences of the economy of the spectacle), as source of romantic inspiration (a resource of a different, supposedly more elevated kind), as anthropomorphic and emotionally-reacting partner ('the angry sea'), or as a lab spectacle where the world is fully yet in a risk-free way reproduced in order to extract more arguments in support of spectatorship.[5] A slightly less fraught word might be the term 'observer', one who does not necessarily partake of the economy of the spectacle, while however retaining the distance thought to be needed for any epistemological judgement. In the remainder of this text, I shall be using both terms – although not interchangeably, since being in the centre of the economy of spectacle remains a position of privilege currently assumed and not easily absorbed by the predominantly epistemological distance that is implied in the term 'observer'.

One of the main tools in approaching this open yet immanent surface is the concept of vulnerability. Such a tool might seem counterintuitive.[6]

---

[3] B Latour, 'On Actor-Network: A Few Clarifications' (1996) 47/4 *Soziale Welt* 369–81 at 370.

[4] G Debord, *Society of the Spectacle* (London, Soul Bay Press 2009) at 4.

[5] J Igoe, 'The Spectacle of Nature in the Global Economy of Appearances: Anthropological Engagements with the Spectacular Mediations of Transnational Conservation' (2010) 30/4 *Critique of Anthropology* 375–97.

[6] See A Grear, 'The Vulnerable Living Order: Human Rights and the Environment in a Critical and Philosophical Perspective' (2011) 2/1 *Journal of Human Rights and the Environment* 23–44.

If anything, vulnerability is a weakness, not a strength. It hardly augurs well for the new semantics. Let me clarify, however, one thing: there is nothing soothing about this new surface and about the way in which this surface presents itself. The surface is precipitous, anti-gravitational, whirling. And the actant/spectator faced with it has not time or space from which to observe it. Any need for observational distance is swallowed up by the velocity at which the surface moves – from polarisation and dialectics to what Heidegger[7] has famously called being thrown-into the world: a forceful abandonment into the world. This abandonment is full of momentum, surrounded by a world that folds around this throwing. There is no screen to hide behind, no distance afforded by epistemology, no negotiating moments of discourse. The surface is pure ontology, inescapably filled with a brutality of continuous, uninterrupted presence. This is the reason for which vulnerability is deemed here to be an adequate tool. To acknowledge this fundamentally precarious position of the surface is to acknowledge one's vulnerability. Further, acknowledging one's personal vulnerability is a condition of limits. One gets to accept the fact that one has limits and therefore limitations. It is precisely the awareness of the latter that is lacking in environmental legal discussions.

Vulnerability is understood here as the condition of finding oneself in a space in which the inevitability of being thrown-in and exposed to the draughts of the world becomes evident to the one who is thrown in. This ('thrown-in') 'one' might be a human being of 'the golden mean type', namely the human praised by Enlightenment as rational (I wonder where these humans are hiding); or a human that still grapples with her humanity, trying to find ways of delimiting herself in a technological or natural milieu; or even a non-human, an animal, plant or inorganic matter whose consciousness (or lack of) has never managed to place all the demands under which humans currently find themselves. Vulnerability does not require human consciousness but a mere instinct of risk of potential harm. It unfolds in an open yet immanent space, situated right *in the middle* (of the surface, of the world, of the body), folded amongst the flesh of an infinite yet immanent plane, yet precipitously falling. Not in a theological sense, although the latter is inspired precisely by the acknowledgement of vulnerability. Unlike theology, here there has been no one who pushed us, no one who gave us the choice – even that of no choice. Here, there is no distance between our substance and that of the

[7] M Heidegger, *Being and Time* (New York, Harper and Row 1962).

divine. We are all one, according to Spinoza.[8] There has never been an apple moment with its soothing quality of origin. Here, the body woke up in the middle of its fall: apples and other fruit, animals, tools, cities, plants, machines, cyborgs, ecosystems, the earth – all form part of the same gravitational movement, itself absorbed by its own force and eventually collapsing. A body (whether it is human, animal, technological, linguistic, inanimate) finds itself in the middle of this, moving and becoming-other-than itself, trying to grapple with its condition shared by everything around it. On the plane of immanence, vulnerability occupies the space of exposure that goes beyond epistemology and beyond the Heideggerian distinction between the ontic and the ontological.[9] We can no longer accept levels of being – or this kind of thought, a staggered ontology that reinforces categories and hierarchies. I want to begin on a clean slate where the surface is *all that there is*. In that sense, vulnerability is a moment of pure ontology, a moment of doubling-up, where the vulnerable (that is, everything) is present in its own vulnerability as its defining condition of presence. As soon as one is present, one is vulnerable.

The space of vulnerability has grave repercussions for the way we understand ecological processes (in the broadest possible sense that includes as much the ecosystemic as the cultural) and the position of environmental law with regards to them. First, it reveals an incontestable exposure to the world. No matter how well protected humanity feels behind its technological armour, it remains exposed to the flow that surrounds it, be this in the form of natural disasters, autopoietically-produced technological risks, human-caused accidents or hybrid forms of life between the natural and the artificial. Second, the space of vulnerability exposes the futility of looking for a centre in the way life, society or anything else is structured. Indeed, the middle is not the centre – if anything, the middle makes the centre collapse. In that sense, the debate between anthropocentric and ecocentric is an exercise in old semantics. And third, it exposes the illusion of hierarchical control believed to be yielded by the law in relation to nature. To begin in the middle is to find oneself in a horizontal plane where no transcendent summit is to be seen and no *über*perspective can be assumed by anything, least of all the law.

On this shifting plane of immanence, I would like to locate an attempt to reconceptualise environmental law in its new becoming, what I would

---

[8] B Spinoza, *Ethics* (Translation, GHR Parkinson) (Oxford, Oxford University Press 2000).

[9] Heidegger (n 7).

call *critical environmental law*. Briefly, critical environmental law's main theoretical task is to reconcile the observer with the actant, namely the epistemology of observation and the ontology of being in the middle. There is no simple dialectic between observation and participation, observer and actant or even spectator and actor: there is no good guy/bad guy here – neither of them corresponds directly to issues of centre and middle, and certainly no synthesis can rescue us from the distinction. At work here is an attempt at weaving the two in strands that retain their individuality yet work together in order for the middle to take up the expansive immanence of the distance that is traditionally reserved for observation. This is particularly important in the face of the absence of boundaries between traditional categories such as human/animal, artificial/natural and so on. If this difference no longer exists, then how can critique be conceptualised? Can one move away from representations and embrace the materiality of the environment? In order to answer these questions, I follow a broadly Deleuzian understanding of the connection between the human and the environmental, on which I build a theoretical sketch of critical environmental law that is inspired by a productive conflict between Luhmann and Deleuze. Thus, in the following section, I begin, brutally in the middle – offering a broken text dealing with its own vulnerability.

After situating the concept of vulnerability in the middle, I look further afield, to the open plane of immanence where there is no differentiation between human/natural/artificial. This is what, in section 3, I call 'open ecology'. In section 4, environmental law is formulated in its paradoxical nature of both *nomos* and *logos*, namely, both as an invitation to disciplinary and ontological openness and as a call to remain within existing legal structures. This is theorised in section 5 as the crisis of environmental law, namely the confluence of critique as demarcation/boundary/crossing over and critique as problematisation of concept. In the final section, the findings are distilled in four critical environmental positions that set the stage for the construction of a critical environmental law.

## 2. TO BEGIN IN THE MIDDLE

To begin in the middle is to denounce the possibility of a centre. The middle does not allow for a perspective that calls itself an origin and from which all is centrally, panoptically surveyed. The middle is not the centre. In fact, it is not even a beginning, properly speaking. It is a movement along movement, something equivalent to stepping out of a door and being carried away by a crowd irresistibly pulling in this and that direction. One finds oneself right in the middle of things, surrounded by a movement that cannot be controlled: 'The middle is by no means an average; on the contrary, it is where things pick up speed'.[10] To begin in the middle (just as to relinquish control) is at the same time debilitating and liberating.

When Deleuze and Guattari emphasise precisely this space of the middle as the point from which to begin, they revolt against the habitual conceptualisation of beginning with its panoply of origin, centre and boundary. Deleuze and Guattari use the example of the grass in order to describe the process. Just as the grass has no one root, central part, or limits to its expansion, in the same way to begin in the middle is to find oneself folded between the multiplicity of the world without a discernible origin, a specific centre or determined territorial limits. To be thrown into the mobile multiplicity of the grass is to follow the blades waving in the wind: one loses one's origin, one's preconceived ideas of location and destination, one's belief in the importance of the centre. One is lost in a horizontal plane of movement, and on this plane one begins by ebbing and flowing between strength and vulnerability.

Grass is opposed to the tree with its defined root, trunk and volume. As Deleuze and Guattari write, 'arborescent systems are hierarchical systems with centers of significance and subjectification'.[11] For this reason they urge, 'make rhizomes, not roots, never plant! Don't sow, grow offshoots!'[12] Offshoots and rhizomes are characteristics of the planar mobility with which Deleuze and Guattari describe the world. Rhizomes specifically encapsulate the ideas of horizontal, trans-species, heterogeneous growth that trammels Deleuzian/Guattarian thought, in that they do not constitute a linear, vertical construction, but a surface where any modulation is absorbed, closed in and eventually spread in lake-like smoothness. However, even radical rhizomes have been routinely fetishised in the literature as being the way to guarantee openness, flexibility and contingency. This marginalises the fact that rhizomes can also be co-opted, overcoded and used in ways that go against the very idea of rhizome, although Deleuze and Guattari specifically write that the rhizome cannot be overcoded.[13]

---

[10] Deleuze and Guattari (n 2) 28.

[11] Ibid 16.

[12] Ibid 24.

[13] M Michulak, 'The Rhizomatics of Domination' in B Herzogenrath (ed.), *An [Un]Likely Alliance: Thinking Environment[s] with Deleuze|Guattari* (Newcastle upon Tyne, Cambridge Scholars 2008).

The above is an example of the complexity of the middle: neither necessarily 'good' or 'bad', positive or problematic, the space in the middle is a space of struggle – in this case, against origins, boundaries, centres. Even better, the space in the middle is a space of encounters with other bodies, a space in which one's body affects and is affected by other bodies, and within this movement of affect, the body is deemed 'good' or 'bad'. It is not a space of prejudgement, of secure values, of fixed constructions. Rather, the space in the middle is precisely *in the middle*: neither this nor that side; but then again, the middle is not a boundary and therefore is not flanked by sides. It simply is – a movement amidst movement. Likewise, it offers no direction: just as the leaves of grass move with the wind, the space in the middle consists of the *encounter* between the grass and the wind. An encounter for Deleuze pushes the encountered parties off their comfort zone of categories and identities, and throws them in a 'mad becoming'.[14] The grass becomes wind and moves along the wind's breath, the wind becomes grass and spreads itself on the ground: becoming itself is pushed deeper in the middle, as it were. The space in the middle offers no chronology and no external causality: all is interfolded in simultaneity and immanence. The wind becomes the grass, the grass becomes tomorrow's grass – its beginning is in the middle, in the space of here, flailing around its movement.[15] Finally, here there is no difference between spectator and actant. Any ontic counting, categorising or calculating becomes absorbed by an all-embracing, self-absorbing ontological singularity. The spectacle of the environment engulfs the spectators, and the boundary between the two is reduced to a feeble epistemological construction, which, even then, is resisted.

How does this happen? What is it that pushes one from the safety of the spectator to the exposure of the actant? More specifically in relation to humans, what is it in this space that brings them closer to their own responsibility? The answer must be sought in the way that vulnerability becomes apparent. To be in the middle is to become *aware* of one's vulnerability that comes with exposure, abandonment and ultimate

[14] G Deleuze, *Difference and Repetition* (Translation, P Patton) (London, Continuum 2004) at 141.

[15] Deleuze and Guattari (n 2, at 10) use the example of the wasp and the orchid: 'The orchid deterritorializes by forming an image, a tracing of a wasp; but the wasp reterritorializes on that image. The wasp is nevertheless deterritorialized, becoming a piece in the orchid's reproductive apparatus. But it reterritorializes the orchid by transporting its pollen. Wasp and orchid, as heterogeneous elements, form a rhizome'.

release from the habitual illusions that, crutch-like, enable the construction of 'man' to carry on. In his move away from phenomenological consciousness, Deleuze sides with (and interprets) Spinoza's triple illusion of consciousness: the illusion of final causes, of free decrees and the theological illusion.[16] In other words, the illusion of taking effects as causes, the illusion of imagining oneself as the origin, and the illusion of a transcendental point of volition and judgement. Denuded from these foundational illusions, one is also deprived of consciousness itself, for 'consciousness is inseparable from the triple illusion that *constitutes* it ... Consciousness is only a dream with one's eyes open'.[17] So, vulnerability is the moment where the spectator takes herself in, in her full actant dimension, and when the actant swallows up the spectacle of herself in the world. It is the moment of awareness of loss of individual consciousness that consequently requires no distance between the position of observation and the world. Vulnerability is a glimpse at the annihilation of the self through the immense spreading of the self on the open surface of the plane of immanence.

But then, can one be in the middle *and* carry on dreaming with one's eyes open? Can one remain a spectator while observing oneself acting out in the middle? Can consciousness be reconciled with the plane of immanence? Is agency a possibility when one is 'thrown-in'? I attempt an answer to these questions in the following sections, since they lie in the foundation of the tension between, on the one hand the plane of immanence – namely the univocity of substance (be this human, animal, plant, inorganic) – and on the other, the need for assuming a *special* responsibility for the ecological predicament as human beings. It is, in other words, the question of distinguishing without divorcing the human from the rest of the world. The tension focuses on this: can we retain consciousness yet abandon its privileged position? Can we assume responsibility without assuming a central position? Divorcing is much easier than mere distinguishing: it is reassuring to divorce the here from the there, that is, to carve a boundary on the surface of the earth and then position oneself in such a way as to observe a chunk of it while leaving another chunk as blind spot behind. This might well be arbitrary. Every distinction between the observed and the non-observed is to some extent arbitrary. But isn't this what epistemology commands us to do? Thus, say that one chooses to observe how environmental law operates at an EU

---

[16] G Deleuze, *Spinoza: Practical Philosophy* (Translation, R Hurley) (San Francisco, City Light Books 1988) at 20.

[17] Ibid.

level with regards to specific atmospheric pollution. This distinction is performed in full awareness of the fact that it entails an exclusion of all other environmental legal issues. But one carries on with it, in a dreamy consciousness of the exclusion and in solid determination to pursue the chosen focus. One learns to live with one's illusions. It might make life bearable, research less chaotic, multitasking less pernicious. It also makes one feel less exposed, vulnerable and abandoned to the movement of the grass. One might even reserve a space for illusions, a bit like a circus corner designated for tricks and hallucinations. This is not to be harshly judged as inadequate or weak. Faced with complexity, one needs patience. Illusions are necessary, and indeed serve a useful purpose, provided that the grander schema in which illusions are contextualised becomes clear. In this case, the schema refers to the tension between spectatorship and participation, observer and actant, epistemology and ontology. And on another level, it refers to that fundamental distinction between divorce and distinction. The difference is that the latter works across both continuum and rupture – that is, across both identity and difference – and can therefore accommodate a plane of immanence that consists of singular differences, as I show below. But how to achieve this space of distinction within the power struggle of human supremacy versus everything else?

The tension is, once again, solved by the entry of vulnerability. Its impact is not direct, in the sense of solution, but indirect, in the sense that no solution is offered. Vulnerability is too forceful to be easily accommodated. The brutality of the exposure is a singularity unto itself, an all-consuming space of relentless, autopoietically-generated horizontality. The awareness of being thrown-in, the exposure that comes along with the realisation of being singular, produces the angst of, to recall Maurice Blanchot,[18] an acephalous community, namely a society that is no longer sovereign of itself, nor can it nominate a sovereign head with any plausible certainty and permanence. Thinking is no longer directional but open, brutally lost, vulnerably liberated. Especially at an institutional level, where law operates, the space of differentiated thinking is well upon us in every single field, and above all in the fields of interest here, namely environmental law and ecological processes. Environmental legal thought has been exposed to the coruscating atmosphere of horizontality, and while the interest of society might be to maintain the illusion of centre, hierarchy, direction ('arborescent systems'), the evolution of

[18] M Blanchot, *The Unavowable Community* (Translator P Joris) (New York, Station Hill Press 1988).

environmental legal thought is radically and irreversibly exposed to the space in the middle.

Still, this is not a straightforward case of urging things to change. Institutional thought remains conditioned by the resilience of its illusions. Institutional self-descriptions keep on legitimising themselves on the basis of such convenient constructions as consensus, communication, dissent.[19] Environmental law hides behind its legal categorisation and justifies its operations on the basis of its legal nature. This means that, however *sui generis*, environmental law desires to remain law, takes refuge in the grand authority that law still (thinks that it) yields. This places law both away from nature, hierarchically 'superior' to it, and apart from the crisis of the acephalous society. Thus – however much criticised in law – authority, categories, hierarchy remain relevant. So, environmental law exemplifies the paradox of, on the one hand, ontological vulnerability, and on the other, epistemological defence. This is the difference between description and prescription: the latter observes things as they are/appear and tries to speak on the level of the 'system', while the former urges things to change towards a 'better' reality. In the following sections, a combination of the two is attempted.

## 3. THE OPEN ECOLOGY OF ENVIRONMENTAL LAW

For the past few decades, environmental law and ecology have been going through a co-evolution of sorts.[20] This connection is now being redefined, not least because both environmental law and ecology themselves are in the process of redefinition. While ecology has been traditionally associated with ecosystemic principles and management methods, its remit has recently been significantly modified. It is now understood that ecology cannot be taken to mean simply ecosystemic management. (Indeed, Herzogenrath[21] following Deleuze and Guattari,

---

[19] See A Philippopoulos-Mihalopoulos, *Niklas Luhmann: Law, Justice, Society* (London, Routledge 2009) for an analysis of the connection between institutional self- descriptions and illusions of identity.

[20] R Brooks, J Ross and V Ross, *Law and Ecology: The Rise of the Ecosystem Regime* (Aldershot, Ashgate 2003); NY Turgut, 'The Influence of Ecology on Environmental Law: Challenges to the Concept of Traditional Law' (2008) 10/2 *Environmental Law Review* 112–30.

[21] B Herzogenrath (ed.), *Deleuze/Guattari and Ecology* (Basingstoke, Palgrave Macmillan 2009).

replaces ecology with geophilosophy.)[22] Nor, however, can ecology be seen merely as being the scientific branch of the theoretical approach that has come to be known as 'deep ecology'.[23] The definition of ecology employed here is both broader and more pragmatic than the above. At the same time, it is ethically and politically positioned, and also removed from traditional semantic structures of ecocentricity, homeostasis, natural equilibrium and so on. Indeed, when in the 1980s Guattari presented us with his three ecologies,[24] namely a 'mental, a natural and a cultural ecology', he envisaged nothing less than an ethicopolitical articulation of the connections between subjectivity, biosphere and society.[25] Amongst these connections, and specifically in the movement between them, I attempt to locate what I call *critical environmental law*, namely an environmental law that exerts a radical critique of traditional legal and ecological foundations, while proposing in their stead a new, mobile, material and acentric environmental legal approach.

This wider new ecology, which I would call *open ecology*, combines the natural, the human, the artificial, the legal, the scientific, the political, the economic and so on, on a plane of contingency and fluid boundaries.[26] This 'and so on' is precisely the space of openness and simultaneous closure of the concept. 'And so on' arrests the future in a way that excludes waiting. The future is always already here, included in an open ecology of no boundaries. Outer space, hybrid technohumans, technologically manipulated meteorological phenomena, 'and so on' are all grounded on the materiality of here, itself open to any definition of materiality that may come from over there. In that sense, open ecology is entirely closed. Any transcending movement is inscribed within, in the recesses of the unknowable here. There is nothing that is not, actually or

---

22 See also A Papuzinski, 'The Idea of Philosophy Versus Eco-Philosophy' (2009) 4/1 *Problems of Sustainable Development* 51–59.

23 A Welchman, ' Deleuze and Deep Ecology' in Herzogenrath (n 13).

24 F Guattari, *Three Ecologies* (Translation, I Pindar and P Sutto) (New Brunswick, NJ, Athlone 2000) at 20.

25 G Genosko, 'Subjectivity and Art in Guattari's The Three Ecologies' in Herzogenrath (n 21).

26 Although such inclusive ideas have been successfully rehearsed in the wider ecological debate (e.g., WHB Baarschers, *Eco-facts and Eco-fiction: Understanding the Environmental Debate* (London, Routledge 1996) at 14, who equates ecology with 'an attitude, a world-view'), there is a difference between a vaguely or even contextualised ethical/ecological view, and the kind of ecology that combines the material and the conceptual without distinction and traversed by continuous movement. For two excellent contributions to the connection between Deleuze and ecology, see Herzogenrath (nn 13 and 21).

potentially, included in open ecology. And nothing that is not, actually or potentially, connected to everything else. Barry Commoner's first law of ecology, namely that 'everything is connected to everything else',[27] takes up a new meaning, where everything now *becomes* everything else. This is a processual rather than a value-based ecology, and, to quote Deleuze and Guattari, 'we make no distinction between man and nature: the human essence of nature and the natural essence of man become one within nature in the form of production of industry'.[28] This makes the understanding of ecology legal, just as it makes law ecological. The 'production of industry', that is the various elements that repeat themselves in nature and humanity in the form of processes and products is the focus of the connection between law and ecology. Values that have led nowhere successful so far are now replaced by a study of the processes that transcend the usual dichotomy between human/natural. By merely focusing on ecosystemic processes and their management, the construction of 'man' retains its centrality, however included it is supposed to be in the ecosystem itself. To differentiate between the human and the natural/ecosystemic on the basis of category is part of the old semantics of the centre, where various approaches vied with each other on who will get there first. But to begin in the middle is tantamount to abandoning the requisite vocabulary for such categorical distinctions. Each thing is different in absolute, an individual singularity repeating itself in difference.[29] Singularity is folded in the universal without, however, losing its absoluteness. Every human is different, just as every animal, plant, pollution particle or artificial limb is different. Their singularity is folded in the plane of an immanence that brings things together while differentiating them. It is a difference that makes a difference and can never lead to clusterings of categories. This plane can be called life, matter, Nature or indeed, Earth. Thus Deleuze and Guattari argued that '[the Earth] is not one element among others but rather brings together all the elements within a single embrace, using one or another of

[27] B Commoner, *The Closing Circle: Nature, Man, and Technology* (New York, Knopf 1971) at 33.

[28] G Deleuze and F Guattari, *Anti-Oedipus: Capitalism and Schizophrenia* (Translation, R Hurley et al.) (Minnesota, University of Minnesota Press 1986) at 4.

[29] Deleuze (n 14) has written specifically on the connection between the two, where repetition can only be 'true' repetition if it is difference. In the case of the law, this can only happen when we move from the 'particular' (which subsumes repetition to the same) to the 'singular' ('not add a second or third time to the first, but carry the first to the "nth" power': at 2).

them to deterritorialize territory'.[30] But this is the coup of the present approach: not only unity but radical difference on a plane of immanence where everything is infinitely folded. Difference, but not divorce. Open ecology is a closed aggregation of infinite openness, a unity that cannot be seen as a unity because it includes the space from which any observer is to be gazing at it. Open ecology is a unity of radical difference and a difference that breaks unity into luminous, self-absorbing singularities. Further, the present approach not only accepts difference, but also conflict: 'to deterritorialize territory' means to destroy, to flatten out, to smoothen up space, to abandon cities to the jungle, bodies to the ground, fishstock to vessels, vessels to tsunamis: in short, continuously to become-other. There is conflict in movement, for moving always displaces others and constantly renders the organised, structured land a thing of the order of the earth, a smooth, absorbing surface.

So, difference exists. Indeed, only difference exists. Even the Earth which is seen as 'the single embrace' is an open whole, its immanence an infinite plane that is nothing but a snapshot of connections: 'as the production of the diverse [it] can only be an infinite sum, that is, a sum which does not totalize its own elements'.[31] Thus, to say that humans are different from animals/plants/artifice on the basis of essence, consciousness or language constitutes a return to a categorising unity, and as such it goes against everything that this new ecology attempts to do. Essences are means of categorisation, clustering, control. Consciousness is an aggregation of illusions. Language has been abused as the basis for the denial of materiality, the platform on which the material has been delegated to a container or a backdrop. But this new ecology posits an intricate connection between singularity and materiality. Singularity is carved out of individual materiality, of each body. For Deleuze, 'a body can be anything: it can be an animal, a body of sounds, a mind or idea; it can be a linguistic corpus, a social body, a collectivity'.[32] What connects all these bodies is matter. But matter is not a dead weight. On the contrary, 'matter is molecular material, not dead, brute, homogeneous matter, but a matter-movement'.[33] In its movement, matter is generative of both order and chaos. It is supple ('molecular'), folding, heterogeneous. It bears singularities and their encounters with other singularities.

---

[30] G Deleuze and F Guattari, *What is Philosophy?* (Translation, H Tomlinson and G Burchell) (New York, Columbia University Press 1994) at 85.

[31] G Deleuze, *The Logic of Sense* (Translation, M Joughin) (New York, Zone Books 1990) at 266.

[32] Deleuze (n 16) at 127.

[33] Deleuze and Guattari (n 2) at 512.

Matter is pure ontology, unrepresentable, uncontainable, infinite. Matter is the stuff of vulnerability, simultaneously a hiding place and a theatre of exposure, bearing bodies that in their movement affect and are affected by other bodies. Yet, precisely because of matter, there is a difference between humans and other bodies. The difference is one of material thickness, control of affective states and ultimately power. When Spinoza[34] talks about bodies affecting other bodies, he refers to a wider striving (*conatus*) that pertains to all bodies. Striving happens for self-preservation and self-preservation constantly requires more power in order to be able to affect other bodies rather than merely be affected by them. The ultimate goal for Spinoza is a joyful existence, where joy must be understood not as happiness but as a passage to a greater power of acting.[35] It is clear from the way humans have organised their affects that their misapprehended joy (and rather obviously not happiness) has reached vertiginous heights. But this is all happening because of the unacknowledged context. Spinozian affects do not take place in a vacuum. They are always related to the greater, infinite oneness of which they form part. Spinoza calls this 'god', yet the main criticism against him has always been that of atheism.[36] This is because Spinoza's god is nothing more than the infinite substance within which everything floats about, trying to achieve its potential. Spinoza's god is the Deleuzian/Guattarian plane of immanence – that ecological openness that remains infinitely immanent, defined only by and through itself. To put it bluntly, this *here* is all we've got. The way we employ our power is our responsibility. For this reason, the initially peculiar reversal of vulnerability does not discount the importance of power. It is that old paradox, that only through our awareness of our limits and limitations do we become stronger.

## 4. LOCATING ENVIRONMENTAL LAW

Environmental law is thrown in the middle of open ecology. This new predicament renders obsolete several of the constituent elements of environmental law. Yet, the law remains law. Its umbilical credentials are not entirely broken, its illusions not utterly shattered, its vulnerability not totally exposed. Its thrown-ness is cushioned by a resilient layer of procedures, concepts, expectations. We are faced with a paradox, or what

---

34 Spinoza (n 8).

35 Ibid.

36 M Della Rocca, *Spinoza* (London, Routledge 2008).

Beressem calls 'radical paradoxical logic':[37] on the one hand, a law that is moving along the new ecology, situating itself on a plane of difference and dealing with such grand and unheard-of (for the law) concepts such as futurity, genetic materiality, nature or catastrophe – thus flirting with movement and spaces in the middle as no law has done before.[38] On the other hand, a law that still employs habitual legal mechanisms and processes in order to make decisions, reverting always to the foundational moment of crisis, of judgement, of partitioning, thus revives the boundary that cuts across the earth, separating it into us and them.

It would seem, therefore, that environmental law echoes precisely the two forms of law described by Deleuze and Guattari: law and *nomos*. They are both spatially defined, in a way indistinguishable from the material geography in which they operate. Law is the space of the 'legal model', of *logos* and sovereign morality: this is what the authors call 'striated space', namely the space of boundaries, enclosures and exclusions. *Nomos*, on the other hand, is associated with 'smooth space', namely a space that is unlike 'Euclid's striated space'; rather, it is a haptic space 'wedded to a very particular type of multiplicity: nonmetric, acentered, rhizomatic multiplicities that occupy space without "counting" it'.[39] While striated space is the space of the law of the *polis* characterised by points, boundaries and zones, smooth space is that of uncountable, uncalculable *nomos*: 'the *nomos* came to designate the law, but that was originally because it was distribution ... one without division into shares, in a space without borders or enclosure'.[40] The difference is not unlike that between a game of chess and a game of Go:

> in chess it is a question of arranging a closed space for oneself, thus of going from one point to another ... In Go, it is a question of arraying oneself in an open space, of holding space, of maintaining the possibility of springing up at any point ... The 'smooth' space of Go as opposed to the 'striated' space of chess. The *nomos* of Go against the State of chess.[41]

---

[37] H Beressem, 'Structural Couplings: Radical Constructivism and a Deleuzian Ecologics' in Herzogenrath (n 21) at 70.

[38] These concepts appear within the law as legal constructions, always filtered by legal functionality. The fact that they are now part of the legal discourse means that the law has neutralised them sufficiently so that they no longer embody any unpredictable risk for the law.

[39] Deleuze and Guattari (n 2) at 409.

[40] Ibid, 420.

[41] Ibid, 389.

Deleuze in particular has always been fixed on a kind of binarism (although he would not call it so) that contains, on the one side, the idea of law, which came to include the discourse of Human Rights in his later work, and involved the state in its oppressive role as coloniser/land controller; and on the other, *nomos*, which is the distribution of territory along the movement of the nomads, without permanent zoning.

It is, however, problematic to circumvent law altogether (even in the striating sense). The nomic alternative is a reconstruction of a nomadic connection to the territory, where space had not (yet) been appropriated by the state and converted into colonisable, profit-generating land. While this still happens in enclaves of indigenous population, it is not difficult to see that even these enclaves are folded in an overwhelming striating presence of capitalist law.[42] And, even if the presence of the state has in several cases been replaced by a collective and spread-out presence of instruments of control that operate beyond, and indifferently to, state boundaries, the law, in the Deleuzian sense, is literally everywhere, permeating every fissure and fold, and populating the air that one breathes. Law's presence, even if disconnected from the state, is resilient, ever-changing, indeed moving. Environmental law specifically moves along the open ecology, at times co-opting this new thought for its own purposes, and at other times genuinely giving itself up to a constant becoming-other (becoming-science, becoming-politics, becoming-resistance, becoming-tree …) that reasserts its singularity. This paradoxical nature of environmental law, simultaneously law and 'becoming-other-than-law', joining its own fragmentation with that of science, economy, geography, politics, philosophy, demands a radical yet grounded reconceptualisation.

This might come along the lines of Deleuzian *jurisprudence*. Deleuze uses the term creatively, as a contextualised, positive and problem-specific thought located in the middle (in French, *milieu*) of concrete situations. Jurisprudence is a philosophical thought that amounts to praxis, and that facilitates the production of radical encounters between bodies and, consequently, the 'prolonging of singularities'.[43] The positive features of jurisprudence, at least the way I want to contextualise them here, include its spatial specificity, its materiality and its acentricity, in that it does not have a given conceptual centre around which its

---

[42] But see A Brighenti, 'Lines, Barred Lines: Movement, Territory and the Law' (2010) 6/3 *International Journal of Law in Context* 17–27 for the need to resist nomadically.

[43] A Lefebvre, *The Image of Law: Deleuze, Bergson, Spinoza* (Stanford, Stanford University Press, 2008) at 112.

protective role is exercised, but moves along the spatial and material context. In that sense, one can conceive of an environmental jurisprudence as being a form of legal concreteness that dives headlong into the case at hand without preconceived structures as to how the law should act. Every case reconstructs the jurisprudential approach in line with an open ecology of material emplacement. In that sense, Deleuzian jurisprudence puts the material into the Derridean ghost of the undecidable.[44] Indeed, for Derrida, the judge is expected to destroy the law and reconstruct it at the moment of decision, thus refusing to submit to the conservatism of the legal regime while, however, still relying on its force-legitimising structures. The ghost of the undecidable comes before the moment of decision as the doubt that opens up law to justice. Deleuzian jurisprudence demolishes structures in the same way as deconstructive justice, but adds to it the pivotal element of material specificity. An environmental jurisprudence could bring a material ethical orientation capable of resisting political co-optation and disciplinary collapse. Such an ethical orientation, however, cannot come from a habitual employment of environmental ethics.[45] While this remains important, it is also culpable of misguiding expectations that society may have of environmental law. To put it bluntly, environmental law cannot save the planet. What is more, society cannot expect environmental law to *want* to save the planet. As Keith Hirokawa writes, although from a perspective that I would be very reluctant to share, 'deeply held beliefs

---

[44] J Derrida, 'Force of Law: The "Mystical Foundation of Authority"' in D Cornell, M Rosenfeld and DG Carlson (eds) *Deconstruction and the Possibility of Justice* (New York, Routledge, 1992) 1–56.

[45] Innovative applications of environmental ethics do promise, if not a way out, at least a certain way of seeing things differently. This is perhaps the greatest contribution of environmental ethics. On its own, however, it is unable to deal with the kind of increasing uncertainty present in environmental law, not least because of its continuing reliance on a human nature that can be changed from within and according to ecological urgencies. It is characteristic, for example, that Stone's observations on ethics and international environmental law conclude with the following, rather disheartening to the present reader, sentence: 'Ultimately, international environmental ethics may be less about human welfare or even the environment, than about the unavoidable moulding of ideal human character': C Stone, 'Ethics in International Environmental Law' in D Bodansky et al. (eds), *The Oxford Handbook of International Environmental Law* (Oxford: Oxford University Press 2007). If even at the arena of international law, arguably the least human-centred legal discipline, the focus of ethics remains the ideal human character, it seems obvious that the connection between ethics and contemporary concepts of agency is still far away.

alone are ill-equipped to achieve progress in environmental law'.[46] And this is perhaps the crux of the problem and indeed, the main challenge. For, although environmental law is required to position itself amidst the ecology of unbounded disciplines, non-linguistic materiality, human/natural/artificial hybrids and looming ecological disasters; at the same time and after all its disciplinary excursus, environmental law must always return to the employment of a legal language and house itself in courts. This means that, however 'better' jurisprudence may appear in relation to the traditional abstraction of the law, it is steeped in the reality-check of the law itself. The paradox of law looms above all attempts to change things, to prescribe a better law, to take up transformative action: law, and environmental law in particular, remains *simultaneously* open to the infinity of ecology *and* restricted by its legal edifice, itself an ivory tower of closure.

Thus, the task of environmental law is to work both along its connection with this open ecology of disciplinary and ontological fluidity *and* its legal structures, in order to construct a new language (*-logy*) to communicate about this new home (*eco* from the Greek *oikos*, 'home'). Indeed, as existing *logos* (language and rationality), more traditionally structured law is spectacularly discredited when faced with the challenges of environmental law: traditional legal dogmatics collapse before the idiosyncracies of environmental law, bastions of reason are replaced by floats of contingency, and trusted linguistic forms manage to attract vehement critique from feminists, ecologists, spatial theorists, autopoets, corporeal materialists and so on.[47] In the same vein, nature as *oikos* has

---

46 K Hirokawa, 'Some Pragmatic Observations about Radical Critique in Environmental Law' (2002) 21/2 *Stanford Environmental Journal* 224–81 at 227.

47 See indicatively R Verchick, 'Feminism, and Environmental Justice' (2007) 13/2 *Cardozo Journal of Law and Gender* 791–832; D Delaney, *Law and Nature* (Cambridge, Cambridge University Press 2003); J Holder and D McGillivray (eds), *Locality and Identity: Environmental Issues in Law and Society* (Aldershot, Ashgate 1999); D Tarlock, 'Is There a There in Environmental Law?' (2004) 19/2 *Journal of Land Use* 213–54. This is more than a 'Galileo problem' as Cassuto puts it, namely that the environment or indeed environmental law could ever affect a pretence of objectivity. For even when environmental law is forced to affect precisely such a pretence, its affect becomes all too apparent: D Cassuto, 'The Law of Words: Standing, Environment, and Other Contested Terms' (2004) 28 *Harvard Environmental Law Review* [online] 79–128. Available from: http://digitalcommons.pace.edu/cgi/viewcontent.cgi?article=1038&context =lawfaculty (accessed 10 Nov 2014).

been demolished long ago, not so much in its death,[48] but significantly in its moving from a fixed background that 'houses' humanity to a machine that can no longer be differentiated from the human: 'Nature is like an immense Abstract Machine … its pieces are the various assemblages and individuals, each of which groups together an infinity of particles entering into an infinity of more or less interconnected relations'.[49]

Amidst this machine, the challenge for environmental law is multiple. In order to simplify it, I would say that it can all be seen as a striving to avoid the binary distinction between observer and actant, namely critical observation on the one hand and participation on the other. To start with, environmental law has to accommodate its paradox of open ecology (which demands participation) and legal structure (which demands a certain critical distance). This has grave repercussions for the way environmental law conceptualises open ecology: namely, both as the successor of the 'death' of nature, and as the orgiastic breeding of this very corpse in such a way that life simply jumps out of it. This is further complicated by the need for language, which can no longer be 'just' a language (the legal language of the spectator) but rather a performance of wholehearted embracing of materiality (an acting out). For the above reasons, it is not coincidental that environmental law is the most readily available means to drag law as a discipline outside its linguistic ivory tower. Environmental law constitutes the clearest example of law in postmodernity, faced with insurmountable dilemmas that range through cultural relativism to decisions about life and death in a biopolitical context and projected temporalities and the future risk of nanotechnology. As Brooks writes, 'a complete re-examination of current environmental law might… [operate] not only as a preface to environmental law, but as a preface to the understanding of all law'.[50] Environmental law has a responsibility to pull law out of its linguistic closure and to land it on the material, the social, the corporeal, the gendered, the spatial, the animal, the molecular. These are the inhabitants of environmental law's new home: no longer based on the distinction conceptual/material, environmental law becomes one with its ecology, one gesture amongst many others, trying both to define itself and to carry on with its job of protecting its home.

---

[48] C Merchant, *The Death of Nature: Women, Ecology and the Scientific Revolution* (London: Wildwood House 1980).

[49] Deleuze and Guattari (n 2) at 512.

[50] M Brooks, 'A New Agenda for Environmental Law' (1991) 6/2 *Journal of Environmental Law and Litigation* 1–17 at 1.

It is clear from the above that there is a need to push further the theorisation of environmental law along the lines of open ecology. No doubt the relevance of theorisation may not always be immediately apparent, especially since it does not contribute to an assumed 'planet-saving' mission and does not amount to a blueprint for action. Still, the various constructed emergencies have reduced environmental law to being mere reaction rather than thoughtful action, to patching up rather than taking a step back and then throwing itself in, and indeed to a not-so-qualified failure in terms of environmental protection. To a large extent, this is attributed to a lack of critique of the processes and goals of environmental law, which is the focus of the following section. We rarely stop to assess, we have accepted as given a certain traditional legal thinking that remains too positivist for the complexity of environmental issues; we have unquestionably reproduced the human/natural division and even the anthropo/ecocentric distinction; we have looked into environmental law merely as law and not as the idiosyncratic transdisciplinary singularity that it is.

## 5. THE CRISIS OF ENVIRONMENTAL LAW

There is no doubt that environmental law is currently going through an identity crisis, at least in terms of its theoretical underpinning, with all the positive and negative consequences such a crisis might have.[51] The 'crisis' of the section heading, however, refers to something slightly different. It is a crisis of critique, namely a crisis of location of the critical perspective within a plane of immanence where all is connected. It is, once again, the problem of spectatorship versus acting, but this time confronted head-on. Indeed, the whole endeavour of critical environmental law is premised on the concept of critique (of existing legal and social structures) which, at least according to some definitions, requires a distancing from the object of critique, a movement away and indeed higher than, the object – an overview, in short: a transcendence. On a plane of immanence, however, where the connecting event is that of exposed, vulnerable matter shaped in elongated singularities of affect, the critical distance would be a betrayal, a departure from the suffering of the matter for a 'higher', 'safer' sphere of almost theological overview. In this sense, critique would imply that whatever the open ecology might be, the law needs to construct for itself a simplified, legalistic representation.

51 Tarlock (n 47).

A representation does not involve the viewers, who remain distant and able to exert critique from the safety of their distance. An ontological event, however, sweeps up the spectators and situates any critique on the immediacy of the connection between the spectator and the object of critique.

Traditionally, the possibility of a critical environmental law presupposes, first, a boundary between the law and its ecology (a distance), and second, the crossing of the boundary (a transcendence). This is because to be critical entails a departure (from the current situation and indeed the object of critique). It also entails an (illusion of) arrival (to a better situation explicitly or implicitly assumed behind the act of critique). In this movement from one side of the boundary to the other, critique performs its actualising, its *krinein*, its distinguishing, indicating and judging: as Douzinas and Gearey write, '*krinein* means also to cut; critique is a diacritical or cutting force, a critical separation and demarcation'.[52] Thus, a legal critique begins from the current state of the law and crosses the boundary that distinguishes the latter from a 'better' state of the law. In its urge to cross, critique announces a crisis (a judgement, a distinction), a crisis of crossing (critique cannot leave itself out) as well as a crisis than can only be observed through this crossing.

However, one must wonder: is it really the case that for environmental law to be reconfigured as *critical* environmental law, a higher, transcendent perspective must be adopted? This goes to the heart of the legal discipline and its ecology. Niklas Luhmann's theory of autopoiesis may be of succour here as the antilogos to Deleuzian normative claims. This is not because autopoiesis is critical *per se*[53] but because autopoiesis professes always to talk on the level of description of what actually happens rather than on the level of normative prescription on what should be happening. It is indeed the theory of spectatorship, in the sense of distant observation, par excellence. The main concept of autopoietic theory is that law (or politics, religion, economy and so on) is an autopoietic system, namely a moving and amoeba-like sum of (legal, political and so on) processes and operations that evolves by learning from its ecological position amidst the greater openness of society. In this way, the system constructs its own self-description, its location and its boundaries, not unlike the Deleuzian understanding of Earth, as commented earlier. The system, however, is contained by a boundary that

---

52 C Douzinas and A Gearey, *Critical Jurisprudence: The Political Philosophy of Justice* (Oxford, Hart 2005) at 38.

53 Philippopoulos-Mihalopoulos (n 19).

determines interiority and exteriority. Although the system learns from its ecology (it is cognitively open to it), nothing ever crosses the boundary between the system and its ecology: in that sense, an autopoietic system is a closed system that observes, from a safe distance, its 'other'. This other, however, is nowhere to be found but inside the system. Everything is internalised and self-constructed within the system, as an infinite plane of immanence folded within the system – except that this one is placed on an epistemological, systemic level, rather than the ontological level of Deleuze's immanence. Everything in autopoiesis constitutes its own singularity, and indeed believes in it: as far as the system is concerned, there is nothing outside its boundary. Yet, other systems, each one believing in its own absolute singularity, carry on constructing their own world, containing everything that they (actually or latently) understand, meteoric formations swirling in a claustrophobic embrace with themselves. And despite all this, everything is somewhat connected, coupled up and operating in unison. For Luhmann, society is equivalent to the Deleuzian/Guattarian abstract machine which is nothing more and nothing less than the infinite sum of its parts.

There are several points of contact between a Deleuzian and a Luhmannian theory of ecology, including the absence of centre, the turning against the linguistic turn, the dilution of the human in the wider ontological (for Deleuze and Guattari) or epistemological (for Luhmann) plane of couplings, and indeed, the infinity of openness.[54] The question of the boundary, however, seems to be the defining difference between the two approaches. Thus, while for Deleuze there are no boundaries, just folds, for Luhmann, the system is defined by (while it itself defines) its boundary. The question of the boundary is not of just theoretical interest. Crucially, it has profound implications for the kind of theoretical advancements that one can pursue in terms of environmental law. If environmental law operates within a boundary that distinguishes it from politics, science, other law, and even the materiality of ecology, its critical transformation will entail a crossing of this boundary, a transgression to a perceived transcendent amelioration: critical environmental law can happen if current environmental law is suspended. Only in its suspension can environmental law generate the kind of critical distance needed for a truly radical new identity. If, however, environmental law does not operate within a boundary but is simply another plane within the plane, an infinite fold of the Deleuzian plane of immanence, its critical manifestation will not involve either crossing or suspension. Critique for

54 Beressem (n 37).

Deleuze and Guattari is a process of problematisation of the concept under critique. By positing the problem (a new problem or an existing problem posited in a new way), critique concentrates on the concept and the way it is now reformulated. As they write, 'to criticize is only to establish that a concept vanishes when it is thrust into a new milieu, losing some of its components, or acquiring others that transform it'.[55] There is no crossing as such, just a throwing of the concept into a new middle (milieu), replacing its elements with others and constructing its connections with its new milieu in a positive and creative way. The consequences of this approach for critical environmental law do not amount to suspension of existing structures and concepts, as they would with critical autopoiesis, but are an infinite unfolding (in the sense of folding out, continuously folding): 'Unfolding is not the contrary of folding, but follows the fold up to the following fold'[56] of its concepts, a continuous throwing-in-the-middle in order to replace their obsolete components.

The positions of Delueze and Luhmann, however, have more in common than it would actually appear at first impression. Let me attempt to weave them together, in an effort indeed to show that spectator and actant are not necessarily so different. For Luhmann, as far as the system is concerned, the boundary does not exist. The system considers itself infinite, spreading itself on thick layers of universal matter and containing all that is to be contained. Does the universe have a boundary? A system does not know any more than we do. The system's self-construction of its outside is (almost) so fault-proof that there is no difference between construction and 'reality'. To put it provocatively, there is no difference between (Luhmannian) epistemology and (Deleuzian) ontology, between the distance of spectatorship and the immersion of the actant. The boundary dis/appears in the form of the materiality of transgression, of critique itself. For support, surprisingly, I return to Deleuze and Guattari's understanding of critique: 'when philosophers criticize each other it is on the basis of problems and on a plane that is different from theirs and that melt[s] down the old concepts in the way a cannon can be melted down to make new weapons. *It never takes place on the same plane*'.[57] The difference between planes is comparable to a

---

[55] Deleuze and Gautarri (n 30) at 28.

[56] G Deleuze, *The Fold* (Translation, T Conley) (London, Continuum 2006) at 6.

[57] Deleuze and Gautarri (n 30) at 28, added emphasis.

systemic boundary: an invitation to jump over, to transgress, and eventually to construct internally in the system or the concept.[58] Of course, just as with a boundary, the crossing never really takes place. It is always the object of critique that deals with the various critiques around it by internalising them in the form of new components, autopoietically reproducing itself like an infinite fold: as Deleuze writes, 'the problem is not how to finish the fold, but how to continue it, to have it go through the ceiling, how to bring it to infinity'.[59] As such, critique is continuous. There never is a beginning of critique in the sense of a boundary cutting through and separating. Critique can only build on critique in a continuous unfolding. At the same time, however, critique always suspends the concept that it criticises, or to put it more precisely, it reveals that the context in which the concept used to operate successfully so far, is no longer relevant, indeed that it has suspended itself. This paradox, which applies specifically to the concept under critique, is mirrored in the more familiar paradox of critique necessitating both distance from and simultaneous immersion in the folds of its object. In that sense, one can revisit the Deleuzian concept of jurisprudence – as applied, materially specific problem handling – and bring it together with the legal need for maintaining a certain closure in terms of the structures it uses, the language it employs, the houses in which it is housed. Jurisprudence operates as the critique of the crisis of environmental law, an applied, hands-on critique that is both enmeshed with the crisis of environmental law and contextualised in an open ecology of connections.

One thing becomes clear. There is no transcendence, no *über*perspective from which environmental law can perform its critique. Either protending its own plane or crossing its boundary, critical environmental law turns 'against' itself, internalising its fall in scathing but positive critique, because 'those who criticize without creating, those who are content to defend the vanished concept without being able to give it the forces it needs to return to life, are the plague of philosophy'.[60] And if I may add, this is precisely the plague from which environmental law suffers. To realise that there is no opposition between maintaining a distance and immersing oneself is the way critical environmental law needs to conceptualise itself. In practical terms, this is nothing more than

---

[58] This is how I interpret Lefebvre's (n 43 at 279, n 8) statement that for Deleuze critique is always and only external. Absolute externality can only become internalised by the object of critique, if the critique is to be at all meaningful.

[59] Deleuze (n 56) at 39.

[60] Deleuze and Guattari (n 30) at 28.

a realisation of the need to renegotiate given balances in such a way that they can accommodate the paradox. In what follows, and as a conclusion to the discussion as well as an opening of further research, I articulate four positions on critical environmental law that integrate the above observations.

## 6. FOUR CRITICAL ENVIRONMENTAL POSITIONS[61]

### 6.1 Position One: Critical Environmental Law Begins in the Middle

In its critical turning, the law must adopt the same level of vulnerability as its object of protection, namely open ecology in its expansive sense of human/natural/artificial. In being thrown in the middle, environmental law, first, relinquishes the illusions of control of nature; secondly, it abandons the search for a centre; and, thirdly, it takes on the exposure needed in order to create the appropriate conditions for its own materiality. For it is one thing to *talk* about how the law needs to engage with materiality, and another to seek and promote the materiality of the law itself. While the former begins by carving a boundary between itself and the material world, equipping itself to confront it; the latter does not differentiate between the materiality of ecology and that of the law itself, thus bringing forth the mobile matter of the law. As Austin Sarat writes, 'the law is all over':[62] this is a call to claim law's spatial, corporeal, emotional, sensory presence that has been subsumed in the critique of discourse. Critical environmental law can no longer hide behind the usual legal logocentric panoply that insulates it from the influx of materiality. More than other legal disciplines, environmental law is exposed to its own materiality: dealings with scientific thresholds, ecological catastrophes, urban poverty, sick bodies and polluted atmospheres – most recently, the challenge of material inscription of the law within the body

---

[61] I have already offered these four positions from an autopoietic perspective in A Philippopoulos- Mihalopoulos, 'Towards a Critical Environmental Law' in A Philippopoulos-Mihalopoulos (ed.), Law and Ecology: New Environmental Foundations (London, Routledge 2010). The present is an exercise in bridging the two streams of thought as well as fleshing out the constructive conflict between them.

[62] A Sarat, '… "The Law is All Over": Power, Resistance and the Legal Consciousness of the Welfare Poor' (1990) 2/2 *Yale Journal of Law and Humanities* 343–80 at 343.

of genetically modified crops.[63] These are relatively recent encounters for the law and from such encounters law can muster the power to change itself. As Halsey writes, law's encounter with the majestic in nature has the potential of impacting 'in ways not immediately amenable to preassigned or preconceived categories';[64] and, one has to add, while maintaining the structures of legal decisions as the calculations towards justice. This is precisely the meaning of being thrown in the middle – ontologically vulnerable, exposed in the middle, and at the same time taking refuge in existing legal structures.

### 6.2 Position Two: Critical Environmental Law is Moving Along its Open Ecology

The dry dualism between anthropocentric and ecocentric legal protection must be relinquished and the blurring between the 'natural', the 'human' and the 'artificial' must be extended in line with biopolitical readings that point to precisely this kind of fragmentation and reinstitute a posthumanist form of continuum/rupture, referring variously to human/non-human nature, and to the way in which these differences are conceptualised.[65] This has been transposed into ecological thinking, with a pioneering ecofeminism, especially through the writings of Val Plumwood[66] and Carolyn Merchant,[67] whose ecocriticism of pre-existing binarisms has had a measurable effect on feminist theory as a whole. Donna Haraway has famously declared that 'the boundary between human and animal is thoroughly breached'[68] and cyborgs, oncomice and coyotes are posthumanist dimensions of more traditional feminist bodies that transcend the natural/cultural, organic/mechanical, physical/non-physical divides.

---

[63] A Pottage, 'Biotechnology as Environmental Regulation' in Philippopoulos-Mihalopoulos (n 61).

[64] M Halsey, 'Majesty and Monstrosity: Deleuze and the Defence of Nature' in Philippopoulos-Mihalopoulos (n 61) at 283.

[65] See M Foucault, *Abnormal: Lectures at the College De France 1974–1975* (London, Verso 2003); S Ruddick, 'Domesticating Monsters: Cartographies of Difference and the Emancipatory City' in L Lees (ed.), *The Emancipatory City* (London, Sage 2004); R Braidotti, *Transpositions: On Nomadic Ethics* (Cambridge, Polity Press 2006); A Sharpe, 'Structured Like a Monster: Understanding Human Difference Through a Legal Category' (2007) 18/2 *Law and Critique* 207–28.

[66] V Plumwood, *Feminism and the Mastery of Nature* (London, Routledge 1993).

[67] Merchant (n 48).

[68] D Haraway, *The Haraway Reader* (London, Routledge 2004) at 32.

Bruno Latour talks about 'hybrid networks'[69] between social, informational and ecological systems, and the 'pluriverse' consisting of collectivities of humans and non-humans that redefine democracy as something that can be found either side of the boundary. Katherine Hayles's digital subjectivity[70] is built on a discontinued and inherently unpredictable conception of the human. Anna Grear's new corporate identity departs from the technocratic and augurs new conceptual spaces of understanding responsibility and vulnerability, as well as this new, perennially perplexing lack of distinction between the human and the non-human.[71] In a critical environmental legal perspective, there is no question on whether trees should have standing.[72] The question, rather, becomes one of *nomic* law, that is, a legal regime beyond the property regime, where the singularity of the tree is equivalent to the singularity of the cyborg. The way the law treats these is through a re-articulation, every time, of the jurisprudential problem rather than through the reliance on immobile rules about standing and legal harm and interest. To mention an example: it is not an exaggeration to say that environmental law's uncontested reliance on the concept of harm is one of its foundational problems. A critical environmental legal task is to continue with disengaging legal action from harm, thereby acknowledging the interconnectedness of the open ecology. In more general terms, critical environmental law is expected to move away from the traditional articulation of the legal problem in terms of existing structure/new problems, and turn against itself, in a jurisprudential critique, indeed an immanent critique, of its very own adherence to structure. The result is not the absence of structure, but a structure that does not stand apart from the wider open ecology.

---

[69] B Latour, *We Have Never Been Modern* (Translation, C Porter) (Cambridge, MA, Harvard University Press 1993); B Latour, *Politics of Nature* (Cambridge, MA, Harvard University Press 2004).

[70] K Hayles, *My Mother Was a Computer: Digital Subjects and Literary Texts* (Chicago, Chicago University Press 2005).

[71] A Grear, *Redirecting Human Rights: Facing the Challenge of Corporate Legal Humanity* (Basingstoke, Palgrave Macmillan 2010).

[72] CD Stone, *Should Trees Have Standing? Toward Legal Rights for Natural Objects* (Los Altos, William Kaufmann 1974).

## 6.3 Position Three: Critical Environmental Law is Singular

Singularity is difference. The seemingly closed unity of differences is an open sum of connections, ridden with conflicts, displacements, movement. Although this goes against the usual starting point of the ecological discourse in the sense of ecosystemic or narrowly ecological unity, along with a unifying centre and a homeostatic goal,[73] it is important to reconceptualise environmental law in its singularity as a body (of discourse) that affects and is affected by other bodies (in every sense). Singularity, however, is not uniformity. Environmental law must be thought of critically, as a multilayered, global, fragmented discipline, characterised by links that follow the transboundary nature of pollution. Differentiated velocities of connections, however, make this a complex affair: the connection, for example, between law and science is one of progressive convergence of very different temporalities (science is faster, law takes longer to deliberate) and as such there can be no facile impression of direct influence and control. To be a singularity amidst a plane of other singularities entails a radical *acentricity* and an understanding that various perspectives operate on the same concept at the same time, with different and not necessarily predictable effects on it. This is more than a call for the inclusion of uncertainty in environmental law. Rather, it is, once again, an invitation to a jurisprudential (in the Deleuzian sense) approach to environmental issues, namely a praxis-oriented, spatially specific, material approach that considers every problem in its singularity. While this sounds commonplace, it is actually radical. Its radicality lies in the constant destruction of existing structures in order to reconstruct them, every time once, every time as first time, in the face of the demands of ontological vulnerability.

## 6.4 Position Four: Critical Environmental Law is Immanent

The overly prescriptive nature of environmental discourse, even in its legal theoretical aspect, urges us towards a rushed search for short-term solutions that misunderstands the complexity of ecological protection. An immanent environmental law acknowledges the necessity of linking to existing law (*logos*), while opening up to the possibility of nomad law (*nomos*). This is a combination of acknowledgment of limitations and of

[73] H Odum, *Environment, Power, and Society* (New York, John Wiley and Sons. 1971); I Blühdorn, *Post-ecologist Politics* (London, Routledge 2000).

potential for a differential evolution. Immanence also precludes transcendence: law cannot assume the higher ground in order to judge (for there is no such thing as higher ground), and critical environmental law cannot rely on distance in order to perform its critique. On the contrary, the critical distance is devoured by the folds of immersion, at the same time as immersion is carried further away on a different critical plane. In this relentless ebb and flow, critical environmental law constantly becomes-other, is thrown into new middles and loses old components. Just as environmental law cannot refrain from being law, it also cannot refrain from *not* being law. Its identity crisis offers the perfect platform for its continuous becoming-other, becoming-science, becoming-animal, becoming-economy, becoming-Earth. Its vulnerability, beyond representation, is embodied in environmental law's continuous 'throwing-in'.[74] Environmental law has the potential of leaving representation behind, of throwing itself in the ontological event of exposure: no longer a law that only deals with discourses, but a law that violently unearths the materiality of the discourse; further, that unearths the matter beyond discourse; even further, that exposes its own materiality on the same plane as the materiality of its object. Critical environmental law is a law frank about its material presence: through its own folds of embodiment, critical environmental law spreads the body of its judgement across the plane of immanence.

[74] See also A Philippopoulos-Mihalopoulos '"... The Sound of a Breaking String": Critical Environmental Law and Ontological Vulnerability' (2011) 2/1 *Journal of Human Rights and the Environment* 5–22.

# 3. Reflections on the relationship between environmental regulation, human rights and beyond – with Heidegger

**Margherita Pieraccini**

## 1. INTRODUCTION

> Let us labour to demonstrate how we must become aware of ourselves.[1]
>
> No-one yet has made the crossing from nature to society, or vice versa, and no-one ever will. There is no such boundary to be crossed.[2]

The separation of human rights from environmental protection has been questioned extensively in the legal literature ever since the connections between the two were first made by Principle 1 of the Stockholm Declaration 1972 on the Human Environment.[3] A variety of scholarly approaches have been proposed over the years to understand the ways in which the borders of the human rights rubric are, can or should be redrawn to incorporate environmental considerations. The most popular options include strengthening procedural environmental rights, creating a new substantive right to a satisfactory environment or reinterpreting

[1] M Mauss, 'A Category of the Human Mind: The Notion of Person; the Notion of Self' in M Carrithers et al. (eds), *The Category of the Person* (New York, Cambridge University Press [1938] 1985) 23.

[2] T Ingold, 'Epilogue: Towards a Politics of Dwelling' (2005) 3/2 *Conservation and Society* 501, 508.

[3] Principle 1 of the Stockholm Declaration 1972 contains this statement: 'Both aspects of man's environment, the natural and the man-made, are essential to his well-being and to the enjoyment of basic human rights – the right to life itself'.

existing human rights to protect the environment.[4] These options are principally the outcome of sophisticated empirical analyses that have enriched our understanding of (human) rights perspectives on environmental protection.

I propose to re-think the debate between human rights and environment by stepping away from these popular rights-based approaches, focusing instead on the interactions of environmental regulation and human rights as two legal fields. I argue that the way in which we are able to think about the possible interactions between these two legal fields is determined by certain philosophical assumptions concerning humanity and environment. As will be seen, once these assumptions are discarded, the task will become one of thinking about a law beyond the human rights/environmental regulation dyad, a law representing the relational unfolding of humanity and environment.

The discussion is articulated here by presenting two successive representations of the legal relationship between human rights and environmental regulation. I term these the oppositional and the interactive views. The oppositional view, discussed in the first part of the chapter, corresponds to the initial phase of the relationship between human rights and public environmental regulation.[5] It focuses on the discordances between a subjective approach to human rights based on rights of given subjects and an approach based on the external, state regulation of a public, de-subjectivised good: the natural environment. The Kantian assumption that lies at the basis of this view, i.e. the subject/object split, positions human rights and environmental regulation in two bounded and conflictual legal realms.

However, some of the conflicts seem to be resolved if we embrace an interactive view, as discussed in the second part of this chapter. This second view is represented by legal and policy developments both in the sphere of human rights and environmental regulation in the past few decades and in related academic discussions. Environmental regulation has been de-centred and has incorporated human rights considerations, while human rights law has been greened, thereby revealing its environmental regulatory

---

[4] There is also a rich literature, possibly originating with Stone, advocating for the development of environmental rights per se detached from any anthropocentric considerations (C Stone, 'Should Trees have Standing? Toward Legal Rights for Natural Objects' (1972) 45 *Southern California Law Review* 450). This chapter, however, will not deal with this body of literature as it will focus on human rights and environmental regulation.

[5] Tort law, as a source of private environmental regulation, is not considered in this chapter.

potential. The 'greening' of human rights law and the 'righting' of environmental law reshape the boundaries and connections between these two legal fields, helping to overcome the oppositional view. However, the interactive view, by drawing most of its inspiration from specific legal examples, constitutes more of an empirical and (often) policy-oriented overcoming of the oppositional view rather than a jurisprudential questioning of the oppositional view's key ontological assumptions. Indeed, the interactive perspective leaves intact the distinction between given human subjects and the external environment upon which the oppositional view is built. As a consequence, human rights and environmental regulation continue to inhabit two separate, albeit now porous, legal spheres.

As the third section of the chapter shows, it is only when, following thought, we abandon Cartesian dualism and focus on the question of the meaning of Being that we are able to think relationally and may start to argue that it can be fruitful (or even necessary) to think of laws beyond the dyad of human rights-environmental regulation. Heidegger is brought into this picture because though not being the only process philosopher (think about Bergson for example), he was the first to explicitly re-open the problem of the Being as distinguished from beings, as we will see below. For our discussion, following Heidegger means that the way forward is no longer to keep debating whether rights should be greened or regulation righted, but to reveal that the two separate legal spheres of human rights and environmental regulation are parts of, and constituent of, a particular way of revealing Being that conceals practices of dwelling and ethics of care. The chapter, accordingly, ends with a call to do more work beyond concepts of rights and regulation in order to uncover other legal approaches to human-environmental relations, opening spaces where a poietic (in the Heideggerian sense of 'bringing-forth') view of law can be disclosed.

## 2. THE OPPOSITIONAL VIEW: RIGHTS VS. REGULATION, HUMAN VS. THE ENVIRONMENT

As Cranston puts it 'human rights belong to man simply because he is a man' and therefore human rights can be universalised as they represent 'something that pertains to all men at all times'.[6] Human rights therefore express and guarantee the pre-existing characteristics and dignity of

[6] M Cranston, *What Are Human Rights?* (London, The Bodley Head 1973) 24, cited in LH Leib, *Human Rights and the Environment: Philosophical, Theoretical and Legal Perspectives* (Leiden, Martinus Nijhoff Publishers 2011).

individual human beings. As is rightly argued by Douzinas and Gearey, however, this assumption is 'part of liberal theory's impoverished view of the subject as a closed and monological entity and of the social bond as an atomocentric collection of individuals whose relations to each other are external'.[7]

To this criticism I would like to add that in the liberal tradition individuals are placed in a position of externality not only in relation to each other but also in relation to the environment. Indeed, within the liberal tradition, human rights are thought of as resulting from *being* (as opposed to *becoming* through socio-ecological relationships) human. Therefore, the environment has no role in the formation of subjects, as human beings are thought of as entering the world as already imbued with essential (in the sense of essence) human traits. From this perspective, the central task of human rights law is to render explicit and to uphold the inherent rights of given human beings. This liberal conceptualisation of the human being is however not an absolute ontological characterisation, but is rooted in Locke's natural law[8] and in Kant's moral philosophy which, influenced by the Cartesian revolution, made individual consciousness the sacred character of the human person. In the early twentieth century, the historical specificity of this view of humanity was considered by the anthropologist Mauss in his seminal lecture[9] on the development of the concept of the person into the individual self. Mauss shows how Kant's *Das Ich* is the modern outcome of a much longer and varied social history, which includes the Roman *persona* (where the person becomes synonymous with the individual and rights are attached to it), passing through the Stoics (where a moral meaning is attached to the already existing juridical one) and acquiring a metaphysical foundation with Christianity.

> From a simple masquerade to the mask, from a 'role'(*personnage*) to a 'person' (*personne*), to a name, to an individual; from the latter to a being possessing metaphysical and moral value; from a moral consciousness to a sacred being; from the latter to a fundamental form of thought and action.[10]

---

7 C Douzinas and A Gearey, *Critical Jurisprudence: The Political Philosophy of Justice* (Oxford, Hart Publishing 2005) 179.

8 It should, however, be recalled that though Locke's definition of property was individualistic, it had some communitarian strains by stressing the importance of utilising the level of resources one needs.

9 Mauss (n 1)

10 Ibid 22.

It is not the validity of Mauss's evolutionary trajectory that should concern us most here, but rather the wider point that can be extrapolated from his lecture: in his words – the way in which the modern philosophical idea of the individual, echoed in international human rights instruments, 'originated and slowly developed over many centuries and through numerous vicissitudes, so that even today it is still imprecise, delicate and fragile, one requiring further elaboration'.[11] The individual subject, then, is not a point of departure but one of interim arrival. Thus, although in the liberal tradition human rights claim to express and uphold inherent universal human properties, in reality they respond to and work with a concept of the individual connected to a particular (and particularly long) Western tradition rooted in Kantian and Cartesian metaphysics.[12]

It is precisely this historically and philosophically constructed concept of the person as an individual subject that is positioned against what is considered to be the principal concern of much environmental regulation, i.e. the protection of the commons[13] from over-exploitation.[14] Hardin's well-known 'tragedy of the commons' is the classical point of departure deployed to justify the existence of much environmental regulation. Following Hardin's argument, in order to avoid the environmental degradation of the commons, external regulation must exist because people, conceived of as an aggregate of individual subjects, cannot by themselves cooperate sustainably. Hardin's argument has considerable resonance with the portrayal of cooperation in classic economic game theory. In the prisoners' dilemma (PD), a non-zero sum game, the rational choice of two players leads both of them to defect because

---

[11] Ibid 1.

[12] It is worth noting that there have been philosophical attempts to provide alternative readings of the person in the Western tradition, for example the work of Kierkegaard as discussed by Assiter (A Assiter, *Kierkegaard, Metaphysics and Political Theory: Unfinished Selves* (London, Continuum International Publishing Group 2009)).

[13] The *commons'* character of the environment is recognised in a variety of legal instruments: from the identification of seabed resources as 'common heritage' in Article 136 of the United Nations Convention of the Law of the Sea 1982, to the loss of biodiversity as an issue of 'common concern' in the Convention on Biological Diversity 1992.

[14] J Brunnée, 'Common Areas, Common Heritage and Common Concerns' in D Bodansky, J Brunnée and E Hey (eds), *Oxford Handbook of International Environmental Law* (Oxford, Oxford University Press 2007); E Louka, *International Environmental Law: Fairness, Effectiveness, and World Order* (Cambridge, Cambridge University Press 2006).

the aim is the maximization of individual payoff, despite the fact that in the long run individual rewards from cooperation would be greater. The underlying assumption of both Hardin and the PD game is that human beings are individual rational appropriators (or polluters) of finite natural resources. Much of traditional 'command and control' environmental regulation operates on these assumptions. 'Command and control' regulation positions the state as the only regulatory solution that commands environmental standards and controls actors through the threat of sanctions. A clear-cut dichotomy between the regulator and the regulated community was created in the early days of modern environmental law, enabled by an understanding of the environment as a natural resource 'out there'.

Similarly to the liberal tradition of human rights law, the initial phases of public environmental law portrayed human beings as pre-formed individual rational subjects acting upon an external common environment existing as a public good, a thing to be safeguarded by the state. However, we must not overlook an important difference between the Kantian liberal tradition of human rights and the Hardinian' self-interest conception of the human being. The difference lies in the role attached to morality: if for Kant rational wills are such only when they follow the moral law, Hardin's self-interested rational appropriators are not directed by moral considerations. However, both views do share the ontological dichotomy between human (individual, rational) subject and the (common, external) environment serving to produce a differentiating effect between these two branches of law, justifying their separation.[15] The aims of human rights and environmental regulation were in fact set far apart; the first concerned with upholding rights inherent in the preconstituted human subjects and with questions of justice, the latter striving to contain these rights to protect common natural resources from overexploitation and concerned with questions of efficiency. This divide was reflected not only in the different legal purposes that human rights and environmental regulation are given within the oppositional view – in the different actors that are positioned at the centre of each (the state in environmental regulation and the individual in human rights) – but also in the different functions assigned to each, with the law on human rights

---

[15] The separation between human rights and environmental regulation is evident for example in relation to property rights/nature conservation law disputes. See for example, *Fisher v English Nature* [2004] EWCA Civ 663 and *Trailer and Marina (Level) Ltd v Secretary of State and English Nature* [2004] EWCA Civ 1580.

having an empowering function, while the law on environmental protection has a controlling function that restricts individual rights to pursue environmental protection framed as a public interest.

Finally, the two legal spheres' engagement with non-law was also different: if human rights law was directly engaged with the political sphere, command and control environmental regulation was relegated to the technocratic field. Human rights law aimed at addressing existing power imbalances, thereby attempting to re-shape existing political contexts, justifying its existence and role on the basis of a-political assumptions (i.e. the inherent characteristics and dignity of given subjects). Command and control regulation by contrast set standards and objectives primarily relying on the discursive power of scientific neutrality and devised enforcement techniques with the goal of achieving effective environmental protection. Consequently, it became impossible to unite environmental and human rights concerns under the same legal umbrella because, due to the basic subject/object split that permeated their existence, both their purpose, functions and engagements with non-law fields (whether politics for human rights or science for environmental regulation) were constructed as separate.

As this section has attempted to demonstrate, ironically it is the philosophical assumption of the human being as a given individual subject, positioned outside an external common environment, that both legal domains subscribed to and which served to produce their legal detachment from each other.

## 3. THE INTERACTIVE VIEW: THE REGULATORY POTENTIAL OF HUMAN RIGHTS LAW AND THE RIGHTING OF ENVIRONMENTAL REGULATION

Observing the development of environmental regulation and human rights law over the past few decades, it seems that the oppositional view portrayed above has long been overcome in favour of a view that stresses the intersections between these two legal fields. This (predominantly) empirical overcoming has two main causes: the de-centring of environmental regulation and the greening of human rights law. Two rich streams of literature have developed around these two causes – critical regulatory scholarship and environmental rights scholarship – both of which have pursued their enquiries independently from each other. After introducing each of these causes and reviewing associated literature, the independence of rights scholarship from regulation scholarship will be questioned following the lead provided by an emerging body of literature[16] that uncovers socio-legal bridges to explore different types of intersections between regulation and rights. A focus on the intersections between rights and regulation reveals the righting of environmental regulation and the transformation of human rights into environmental regulatory instruments. Nevertheless, the ability of this new analytical focus to fully transcend the oppositional view will be questioned in this chapter.

Since the mid-1980s and especially during the 1990s, there have been a series of regulatory reforms in the environmental field (and beyond) that have questioned the validity of classic command and control forms of regulation. Market-based economic instruments have been considered to be an efficient policy option for internalising environmental externalities operating on a decentralised level through their impact on market signals. In the environmental sphere, moving away from command and control towards economic instruments has mostly been visible in the widespread use of 'cap and trade' mechanisms. Cap and trade mechanisms have become a popular regulatory solution in various environmental regimes concerned with commons in order to avoid Hardinian tragedies. Focusing on Europe, fisheries and climate protection are two obvious examples. In relation to fisheries, the Common Fisheries Policy requires the Council of Fisheries Ministers to set an annual cap, known as the total allowable catch, and tradable quotas are then allocated to Member States under a principle known as 'relative stability'. The EU Emission Trading System (Directive 2003/87/EC as amended) operates in a similar manner to reduce greenhouse gas emissions: a cap of total emissions is calculated and companies receive emission allowances that are tradable. The assumption underlying cap and trade is that there is a scarce/fragile environmental common resource out there, fragile due to human overuse (which is, note, based on the same assumption concerning a dichotomous construction of human/environment relationship underlying the oppositional view above), so that a quantitative limit on exploitation must be set determining tolerable use rights (cap), which are then allocated to users and made tradable.

Although economic instruments have often been portrayed as an alternative to command and control and as being a departure from state

---

[16] B Morgan (ed.), *The Intersection of Rights and Regulation: New Directions in Socio-legal Scholarship* (Aldershot, Ashgate 2006); B Morgan, *Water on Tap: Rights and Regulation in the Transnational Governance of Urban Water Services* (Cambridge, Cambridge University Press 2011).

law,[17] this is a misleading representation for two main reasons.[18] First, although regimes operating on cap and trade are highly decentralised, there are other types of economic instruments which remain much more dependent on governmental prescriptions, such as subsidies for carrying out activities that have a positive impact on the environment. For example, agri-environmental measures under the Common Agricultural Policy provide funding to farmers for the provision of environmental services. Secondly, scholars argue that, in certain contexts, environmental regulation can be more effective if command and control and economic instruments are used jointly.[19] Consequently, although economic instruments appear to discard the centrality of the state, the state does not disappear. State control remains one of the forms of control.

Even in environmental regimes primarily based on self-regulation some form of external intervention/control is common. For example, in the EU eco-management and audit scheme, although the evaluation and report of environmental performance are primarily in the hands of the organisation voluntarily participating in the scheme, the initial environmental review, the environmental management system, and the audit procedure are to be verified by an accredited or licensed environmental verifier (Art 4). Also, procedures for the registration of the organisations are to be established by a competent body designed by the member state (Arts 11 and 12) and in case of non-compliance with the Regulation, the competent body can delete or suspend the organisation from the register (Art 15).

We can therefore conclude that although environmental regulation encompasses a variety of measures well beyond traditional command and control, it would be erroneous to portray decentred environmental regulation as being regulation without the state. Regulation scholars who have tracked the decentring of regulation[20] have contested the idea that

---

[17] See for example, S Smith and BH Vos, *Evaluating Economic Instruments for Environmental Policy* (Paris, OECD 1997).

[18] For another critique of this representation see D Sinclair, 'Self-regulation Versus Command and Control? Beyond False Dichotomies' (1997) 19/4 *Law and Policy* 529.

[19] N Gunningham and D Sinclair, 'Regulatory Pluralism: Designing Policy Mixes for Environmental Protection' (1999) 21/1 *Law and Policy* 49–76; G Winter, 'The Climate is No Commodity: Taking Stock of the Emission Trading System' (2010) 22/1 *Journal of Environmental Law* 1–25.

[20] J Black, 'Decentring Regulation: The Role of Regulation and Self-regulation in a "Post-regulatory" World' (2001) 54 *Current Legal Problems* 103; J Black, 'Critical Reflections on Regulation' (2002) 27 *Australian Journal of Legal Philosophy* 1; N Gunningham and P Grabosky with D Sinclair, *Smart*

regulation is hierarchical and tied exclusively to the state – but have not erased the state. The state therefore remains an actor among others, and state law remains one of the normative authorities among others in complex horizontal processes of ordering. Decentred analyses of regulation have pointed out the variety of actors involved in regulation (from the state, to industries, to third parties), the different policy strategies used or recommended for effective regulation and the existence of a wide array of normative fields outside state law with high degrees of regulatory power. The issue of power is in fact a key aspect of a decentred understanding of regulation. In the words of Black, 'this is the recognition that government does not have a monopoly over the exercise of power and control, rather that [power] is fragmented between social actors and between actors and the state'.[21] Yet, by focusing on actors and types of regulation understood as social empirical facts, de-centred regulation has done little to enquire about, let alone to discard, the ontological assumptions concerning humanity and environment that we found at the basis of the oppositional view.

In this sense, decentralised regulation scholarship is in line with legal pluralist analyses that have looked at the interactions of a plurality of legal orders as an empirical, social fact. Legal pluralism of this kind finds its origin in anthropological studies of colonialism, where indigenous forms of law were empirically identified and contrasted to the colonial rule. Since the late 1970s, this dichotomous simplistic picture of legal pluralism has been elaborated upon, and studies have attempted to demonstrate the existence of multiple and interpenetrating legal systems in any given social field, moving across time and space. As early as the work of Moore,[22] the notion of the 'semi-autonomous social field' was employed to draw attention to the connections between the internal workings of an observable domain and its points of articulation and intersections with larger settings. With the concept of 'reglementation', Moore accounted for the interactions between governmental law and

---

*Regulation: Designing Environmental Policy* (Oxford, Oxford University Press 1998); C Scott, 'Regulation in the Age of Governance: The Rise of the Post-regulatory State' in J Jordana and D Levi-Faur (eds), *The Politics of Regulation in the Age of Governance* (Cheltenham, UK and Northampton, MA, Edward Elgar 2004).

[21] Black (2001) (n 20) 108.

[22] SF Moore, *Law as Process: An Anthropological Approach* (London, Routledge and Kegan Paul 1978).

unofficial sites of rule-making and rule-enforcing and moved the geographical focus from colonised to Western societies.[23] As reviewed by Merry in an important piece for the taxonomic development of legal pluralism, the 1980s witnessed the birth of what Merry identifies as the 'new legal pluralism': a geographically expanded pluralism accompanied by a conceptualisation of 'fluid and opaque' links between legal spheres.[24] The seeds of post-modern sensibility are present in this new legal pluralism, since clear-cut power inequalities between two legal orders are rejected and replaced by varied legal connections in a multiplicity of social settings. This logically brings us to the concept of 'interlegality', coined by De Sousa Santos.[25]

Interlegality is, for De Sousa Santos, the 'phenomenological counterpart of legal pluralism'; it describes the contemporary legal porosity we live in and we experience. In fact, what De Sousa Santos identifies as 'post-modern legal pluralism' is, by his own admission, a third phase following the two identified by Merry. The third phase does not reject the view of legal pluralism as an empirical social fact but changes the context from one of infra-state to one of suprastate plurality and power relations: 'global orders coexisting in the world system with both state and infrastate legal orders'.[26]

Issues of power are also considered in other accounts of legal pluralism that focus on actors' strategic appropriation of concepts from other legal systems to fulfil their political claims. This is clear in the writings of legal anthropologists that have looked at processes known as 'forum shopping', where individuals choose among a multiplicity of legal frameworks to better support their claims. Thus, von Benda-Beckmann[27] looked at forum shopping in West Sumatra. Because of the overlap between institutions deriving their authority from *adat* (indigenous) law and from national law, disputants can choose between different institutions, their choice determined by what they perceive to be the most beneficial forum for winning the dispute. More recently, Meinzen-Dick and Pradhan[28] show how disputants over water resources choose among

---

[23] Ibid 18.

[24] SE Merry, 'Legal Pluralism' (1988) 22/5 *Law & Society Review* 869, 874.

[25] B De Sousa Santos, *Towards a New Legal Common Sense* (2nd edn, London, Lexis Nexis Butterworths Tolley 2002).

[26] Ibid 92.

[27] K von Benda-Beckmann, *The Broken Staircase to Consensus: Village Justice and State Courts in Manangkabau* (Dordrecht, Foris 1984).

[28] RS Meinzen-Dick and R Pradhan, 'Legal Pluralism and Dynamic Property Rights' (2002) *CAPRi Working Papers* 22, International Food Policy Research

different legal frameworks to assert their water rights. Power and social relationships among the claimants determine which choice is accepted as legitimate so that rights to water are not rigidly determined but are the outcome of a negotiating process. These writings therefore make agency a key component for understanding the phenomenon of legal pluralism.

After this brief review of this empirical strand of legal pluralism, the point of contacts with decentred scholarship are self-evident, and rest on an understanding of law as an empirical social fact. Decentred frameworks of regulation and plural legal 'constellations', to borrow a term coined by the von Benda-Beckmanns[29] are out there, actors are aware of them, and, sometimes able to forum shop. Moreover, actors and regulatory fields are portrayed as highly dynamic and complex but they always remain wholly encapsulated in a social realm where they are identifiable and distinguishable from the environment.

How can these analyses inform the discussion on human rights law and environmental regulation? Rather than conceiving human rights and environmental law/regulation as bounded and opposed legal manifestations, these analyses can help account for the dynamic relations between these fields and the roles of different actors in appropriating and reconstituting legal discourses within and beyond the official lines. In Europe, an example of this is in the way individuals have strategically sought remedies from the ECtHR where national authorities and courts are not enforcing the environmental law. Although, as is well known, the European Convention on Human Rights (ECHR) does not expressly guarantee a right to a clean/satisfactory environment, the ECtHR has examined a number of complaints by individuals related to a breach of one (or more) of their Convention rights (principally Art 8, Art 2, Art 6 and Art 1 Protocol 1) due to adverse environmental factors. Together with demonstrating the environmental regulatory potential of a human rights instrument, the environmental jurisprudence of the Court is also a demonstration of the way in which human rights law and its institutions are responsive to environmental considerations in line with the Court's understanding of the ECHR as a 'living instrument'.

Yet, the connections between human rights and environmental regulation extend beyond the issues just described. In fact, the greening of human rights is not only an example of forum shopping and of the

---

Institute (IFPRI), at: http://www.ifpri.org/sites/default/files/publications/capriwp 22.pdf.

[29] See for example, F von Benda-Beckmann and K von Benda-Beckman, 'The Dynamics of Change and Continuity in Plural Legal Orders' (2006) 1 *Journal of Legal Pluralism* 44.

environmental regulatory potential of the ECHR law, but also points to other legal pluralist connections between the legal fields of environmental regulation and that of human rights. This is evident if we consider the relationship between the Aarhus Convention 1998 and the ECtHR. The Aarhus Convention 1998 is an environmental law instrument requiring parties to guarantee rights of access to information, public participation in decision-making, and access to justice in environmental matters in accordance with the Convention provisions (Art 1). If the Aarhus Convention mirrors many human rights texts[30] borrowing rights language to legally concretise Principle 10 of the Rio Declaration, it also has an effect on human rights law. In *Taşkin*,[31] the ECtHR explicitly referred to the Aarhus Convention to rule that there had been a breach of Article 8 of the Convention. This is an interesting case for two reasons: first, the Aarhus Convention was invoked even if Turkey was not a party to it; and secondly, a procedural right to environmental information arose not in connection to procedural rights of the ECHR but as prerequisite for the enjoyment of substantive rights of the Convention, thereby blurring the divide between procedural and substantive rights.

By embracing a decentralised/pluralist understanding of law and regulation to understand human rights law and environmental regulation, we cover the same analytical terrain and focus on the same questions considered by scholars who have approached the matter from a rights perspective.[32] The majority of scholarship on human rights and the environment has in fact mushroomed around three main human rights approaches to environmental protection articulated by Anderson in 1996:

[30] DK Anton and D Shelton, *Environmental Protection and Human Rights* (Cambridge, Cambridge University Press 2011) 408.

[31] *Case of Taşkin and Others v Turkey*, App. no. 46117/99, ECHR 2004-X.

[32] See Anton and Shelton (n 30); A Boyle, 'Human Rights and the Environment: A Reassessment' (2009) Working Paper presented at UNEP high-level expert meeting on the New Future of Human Rights and Environment (UNEP); A Boyle and M Anderson, *Human Rights Approaches to Environmental Protection* (Oxford, Oxford University Press 1996); J Ebbesson, 'Participatory and Procedural Rights in Environmental Matters: State of Play' (2009) Working Paper presented at UNEP high-level meeting on the New Future of Human Rights and Environment (UNEP); O Pedersen, 'European Environmental Human Rights and Environmental Rights: A Long Time Coming?' (2008) 21/1 *Georgetown International Environmental Law Review* 73.

(1) the mobilization of existing rights; (2) the reinterpretation of existing rights; and (3) new human rights to environmental protection.[33]

It is fair to admit that there are areas where rights scholarship and environmental regulation/pluralist scholarship go in different directions. A central concern of rights scholarship (not really shared by regulation analyses) is upon the normative validity of a right to a healthy/decent environment as a new human right. For its part, a regulation perspective may put more emphasis upon the way in which environmental laws and regulations are borrowing the concept of rights to improve effectiveness. For example, the climate protection and fisheries regimes discussed above rely on the concept of property rights in order to function. Fishing quotas and emission allowances are in fact forms of property entitlements in environmental resources that serve the goal of un-commoning the environment.[34] As discussed above, the righting of environmental regulation is also visible in procedural environmental law instruments such as the Aarhus Convention. Despite some differences of focus overall, however, both the rights perspective and the regulation/pluralist one recognise human rights law and environmental regulation as being two interactive, porous legal spheres in relation to processes by which human rights law is (or can be) expanded to develop its regulatory potential, and environmental regulation is (or can be) righted. In this sense such approaches seem to overcome the oppositional view and its clear-cut divisions in relation to actors, purposes and engagement with the non-legal spheres. Actors beyond the state enter the regulatory realm, environmental justice considerations become a key point of discussion when assessing regulatory regimes, human rights law is tested for its

---

[33] So much has been written on this topic that it is not necessary to review the literature here. For a recent overview of these approaches, see Anton and Shelton (n 30).

[34] For a detailed analysis of the links between the concept of property rights and fisheries, see R Barnes, *Property Rights and Natural Resources* (Oxford, Hart Publishing 2009). For emission allowances as property rights, see the recent case of the High Court of England and Wales (*Armstrong DLW GMBH v Winnington Networks Ltd* [2012] EWCH10 (Ch)) where the judge, applying the test enunciated by Lord Wilberforce in *NPB v Ainsworth*, considered European Union Allowance as a property right since '[i]t is definable, as being the sum total of rights and entitlements conferred on the holder pursuant to the ETS. It is identifiable by third parties; it has a unique reference number. It is capable of assumption by third parties, as under the ETS, an EUA is transferable. It has permanence and stability, since it continues to exist in a registry account until it is transferred out either for submission or sale and is capable of subsisting from year to year' (para 50).

efficiency as a regulatory instrument for protecting the common environment and the emancipatory character of environmental regulation is recognised in the multiple discussions on participation and deliberation connected, for example, with the Aarhus Convention.

Nevertheless, these discussions do not question the subject/object split that was at the basis of the oppositional view and that produced the legal detachment (of actors, purposes, functions and so on) between environmental regulation and human rights law. Human rights may be greened and regulation righted but the conceptual boundaries of humanity and the environment remain intact, whether we speak of procedural human rights, the regulatory effects of human rights law (due to litigants' forum shopping and/or ECtHR green interpretation of the Convention rights) or even the appropriation of individual property rights concepts by environmental regulation to stimulate processes of un-commoning. For example, much of the academic discussion on procedural environmental rights under Aarhus and ECHR has focused on questions of participation and on the strengths/limitations of proceduralisation in the wider contexts of environmental regulation[35] and human rights law,[36] but not much emphasis has been put on the way procedural rights in environmental matters enable or limit a post-dichotomous view of humanity and the environment. Similarly, the limitations of human rights instruments, such as the ECHR, for addressing environmental matters have once again been considered without questioning the human vs. environment dichotomy.

Again, academic critiques have pointed out that the use of human rights instruments is like 'fitting the potentially round peg of environmental concerns into the square hole of staunchly anthropocentric human rights'[37] and, according to Stallworthy, 'there is an inherent mismatch between the kinds of anthropocentric interests reflected in human rights formulations and protection of the environment'.[38] At the basis of these critiques there is the assumed division between environment and humanity as two separate entities, which is then reflected in the assumptions that human rights law is primarily anthropocentric and environmental regulation ecocentric, that when human rights law is greened it assumes regulatory, controlling functions and that when

[35] M Lee and C Abbott 'The Usual Suspects? Public Participation under the Aarhus Convention' (2003) 66/1 *The Modern Law* Review 80.

[36] Boyle (n 32).

[37] Hamilton, cited in Leib (n 6) 84.

[38] M Stallworthy, *Understanding Environmental Law* (London, Sweet and Maxwell 2008) 57.

environmental regulation is righted it assumes emancipatory functions and takes on board justice considerations. So rights are still about emancipation and recognition of inherent properties of given subjects and regulation about control and ordering. Accordingly, the unquestioned ontological assumption lying at the origins of the separation between human rights and environmental regulation remains intact in the interactive approach to human rights and environmental regulation – thus the distinction between the conceptual value of human rights and environmental regulation is perpetuated. Liberal understandings of human rights are imported to the environmental field and, vice versa, the understanding of the environment as an external resource is imported into human rights law. If the properties of the original constituents of the relationship (humanity and environment) are left intact – in the sense implied here – then the two legal spheres representing them will be conceived of as occupying different, albeit interactive, spaces. This matters because the debate on ways forward now seems to have reached a dead end: either environmental regulation is righted or human rights law greened. This choice is a dead end because it cannot transcend its ontological limitations, i.e. the dyadic understanding of rights and regulation. A different ontological basis therefore may be needed if we want to avoid the dead end. To do so, the next section draws on Heidegger, the philosopher who, in his sustained critique of Western metaphysics and humanism, called for an ontological shift that rejected the subject/object split.

## 4. BEYOND THE RIGHTS/REGULATION DYAD WITH HEIDEGGER'S QUESTIONING OF THE SUBJECT/OBJECT SPLIT

Heidegger's thought is complex. Discussion of it here will necessarily be limited to certain key traits that will serve to critique both the interactive and oppositional views and to open up the possibility for rethinking the legal relationship between environmental regulation and human rights.

For Heidegger, ontological questions should take priority over epistemological ones His life-long interest was in the meaning of Being as distinct from beings (entities). This, as he terms it, 'ontological difference' had been forgotten in Western metaphysics where being has become reduced to what there is, an already-made, finished product, an entity. Being then rather than being about the Being of beings has been relegated to the abstract, whether God in Christian tradition or in the Enlightenment concept of man (or better, of human reason).[39]

For Heidegger, there cannot be access to entities without understanding/disclosing their Being, and this disclosure is, as he puts it, 'factical'. To recover the meaning of Being, the Heidegger of *Being and Time* (BT)[40] starts from the perspective of human experience (*Da-sein*) because *Da-sein* is the being for whom Being is an issue. *Da-sein* helps in revealing, rendering present the Being of beings. This does not happen from a position of externality but by being-in-the-world. Being-in-the world designates the constitution of the Being of *Da-sein*. Importantly:

> we cannot understand by this the objective presence of a corporeal thing [*Korperding*] (the human body) [*Menschnleib*] 'in' a being objectively present. Nor does the term being-in designate a spatial 'in one another' of two things objectively present, any more than the word 'in' primordially means a spatial relation of this kind. 'In' stems from *innan*, to live, *habitare*, to dwell.[41]

Different from the ontological assumptions at the basis of both the oppositional and interactive views discussed above, there is for *Da-sein* no position of exteriority from the world; 'it is not the case that human being "is", and then on top of that has a relation of being to the "world" which it sometimes takes upon itself'.[42] It is this situation of necessary embededdness and relationality (not reducible to ontic characteristics or to Cartesian spatial coordinates) that allows *Da-sein* to discover beings encountered in the environment and to know (unconceal) the world.

In his later philosophy, Heidegger abandons any traces of subjectivity to embrace the question of Being from the perspective of Being itself rather than from that of *Da-sein*, asking how Being unfolds historically. Nevertheless, this philosophical turn (or as he called it, the *Kehre*) is not a change of standpoint, as he himself writes in the 'Letter on Humanism'.[43] There is a high degree of coherence/continuity between Heidegger's earlier and later writings. In the *Letter on Humanism*, in fact,

---

[39] It was with Descartes that for Heidegger the Being became the human subject, turning classical metaphysics into anthropology (C Douzinas, *The End of Human Rights* (Oxford: Hart Publishing 2000) 204).

[40] M Heidegger, *Being and Time* (New York, State University of New York Press [1926] 2010) 20.

[41] Ibid 54.

[42] Ibid 57.

[43] M Heidegger, 'Letter on Humanism' in D Krell (ed.), *Heidegger: Basic Writings* (London, Routledge [1947] 2011).

the *humanitas* of man is not to be understood in terms of some metaphysical substance but is a temporal event (as it was in BT): it unfolds in the movement of life and through this movement it lets beings come into being. The disclosive '*da*' of *Da-sein* remains present in the later Heidegger and it is the *da* that makes it possible for entities to emerge into presence. In the *Letter on Humanism*, Heidegger therefore departs from Sartre's existentialism, which overturned the classical conception that *essentia* preceded *existentia*, by critically pointing out that the differentiation between *essentia* (as essentiality) and *existentia* (as the actuality) remains alive in Sartre's reversal in line with the oblivion of Being. As is well known, to depart from this tradition, Heidegger coins a new term, *Ek-sistence*, as 'ecstatic dwelling in the nearness of Being'.[44]

By following Heidegger's critique of Western metaphysics we can re-think the relationship between human rights and environmental regulation from an ontological view-point. This enquiry starts by uncovering the limits of the ontological assumptions lying at the basis of the liberal tradition of human rights law and environmental regulation that cause the separation between these two legal fields in the first place. As mentioned above, the liberal understanding of human rights finds its historical roots in Locke's natural rights as rights possessed by individuals prior to their involvement in the political community or society.[45] Yet, if for Locke, God remains the source of natural law, with Kant, the human being as a rational and autonomous subject is all that is needed to justify universal human rights. The determination of what it is to be human happens in advance and justifies the possibility of having rights. Heidegger's discussion of *Da-sein* as being-in-the-world is precisely a refutation of: (1) the Lockean possibility of individuals existing as fully formed entities outside any social contexts; and (2) the Kantian subject built on the Cartesian opposition between *res cogitans* and *res extensa*.

As human rights law is based on the idea of a detached substance with inherent properties that forgets the *ek-sistential* condition of being-in-the-world, environmental regulation is based on an understanding of the environment as an external, extended entity: hence their separation is justified. Human rights and environmental regulation as two legal fields are therefore the products of the metaphysical urge of Western philosophy to create the dichotomy between subject and objects, which, in

---

[44] Ibid 167.

[45] Locke's Two Treatises of Government in I Shapiro (ed.), *Two Treatises of Government and a Letter Concerning Toleration by John Locke* (New Haven, CT, Yale University Press 2003).

discussing beings, has forgotten the ontological difference between beings and Being. Rights and regulation, whether in opposition or in interaction, speak about the human and the environment as two entities, without considering the way the meaning of *Da-sein* unfolds (and changes) through its absorption in the environment and, through this engagement, through the '*da*' (which is not a statement of place but of openness), how beings can become present/revealed. It is only when we begin to realise that this way of understanding rights and regulation depends on the way we have come to conceive beings, forgetting the meaning of Being, that we may call for an alternative law that dispenses with assumptions of a priori demarcated subjects.

## 5. CONCLUSION

This chapter has offered a re-focusing of the debate concerning the relationship between rights and regulation. This has been achieved by bringing into the debate on human rights and environment the insights of regulatory and legal pluralist scholarships. This has enabled us to concentrate on the legal relationships between two legal fields, rather than considering the instrumental value of one legal field (human rights law) for the protection of the environment. It has been argued that the initial oppositional phase that positioned human rights and environmental regulation as two opposed realms has long been overcome, and that this is visible in the developments forged both by regulatory/legal pluralist scholars and by 'environmental rights' scholars. These new academic insights, which tend to reflect empirical legal developments, serve to point out the interactions between human rights law and environmental regulation, so that the two legal spheres are now recognised as porous. Nevertheless, even this interactive view operates on the same ontological assumptions that were at the basis of the formation of human rights law and environmental regulation as two antagonistic spheres, i.e. of the oppositional view. If the oppositional view was a product of the dichotomy between subject/object (as ontological assumption) enabling the constitution of human rights and environmental regulation as two separate fields, it follows that, if the same ontological assumption continues to permeate the interactive view, human rights and environmental regulation will continue to exist as separate, albeit interactive, spheres and that the debates on their interaction will continue to offer the same two limited possibilities: the righting of environmental regulation, or the greening of human rights law.

The Heideggerian deconstruction of Western metaphysics offered here has highlighted that the ontological assumption on which both human rights law and environmental regulation rest is a particular manifestation of Cartesian subjectivism due to the current oblivion of Being. By showing that the (separate/separative) existence of human rights law and environmental regulation is only a manifestation of the oblivion of Being, the chapter has opened up a space for academic debate to re-think law poietically, beyond the dichotomy rights/regulation, starting from the realisation that the categories these two legal spheres represent (humanity and environment) are at present reduced to beings, to entities. The task for legal scholars is therefore to produce a new language that houses Being, disentangling itself from the constraints imposed by the tradition of Western metaphysics and able to 'bring-forth' humanity and environment in their Being.

Academic attempts advocating similar paradigm shifts in the law in order to change the relationship between human and nature are to be found in Cullinan's *Wild Law*[46] and more recently Weston and Bollier's *Green Governance*.[47] Both accounts, though very different in their articulations and suggestions, are attempts at changing modern human behaviour towards the natural environment from one of exploitation to a mutually enhancing presence. What the Heideggerian line of thoughts adds to these accounts is an ontological realisation on the *ek-sistential* condition of being-in-the-world and the consideration of the man as the steward of being; something that has inspired philosophers such as Hans Jonas[48] to search beyond rights for an ethical duty to nature.[49] This ontological realisation becomes the foundational basis to rethink law overcoming the dyadic representations of human rights and environmental regulation.

---

[46] C Cullinan, *Wild Law: A Manifesto for Earth Justice* (2nd edn, Totnes, Green Books 2011).

[47] B Weston and D Bollier, *Green Governance: Ecological Survival, Human Rights and the Law of the Commons* (Cambridge, Cambridge University Press 2013).

[48] H Jonas, *The Imperative of Responsibility* (Chicago, University of Chicago 1984).

[49] For a recent consideration of Jonas's philosopher in the environmental field and more specifically in a discussion on corporate behaviour and climate change, see S Wheeler, 'Climate Change, Hans Jonas and Indirect Investors' (2012) 3 *Journal of Human Rights and the Environment* 92.

# 4. Radicalism and conservatism in environmental law

**Sean Coyle**

The nature of law has long been a central question of the Western philosophical tradition. For much of its history, the question formed part of a broader inquiry into the human condition, and the means by which human beings, through collective effort, might realize a good and virtuous form of life. With the introduction of Christianity into Europe, these inquiries took on a different significance. Natural law theories sought to explain the entire basis of governance and law through contemplation of a divinely created natural order of which human beings are but one part. It is often remarked that the present appetite for human rights is explicable by the need to fill the gap left behind by the departure of natural law ideas in the West: that human rights doctrines are the pale, secularized echo of no-longer-held beliefs in a normative order immanent in nature. But the natural law theories of the major Christian writers centred upon ontological questions concerning the fundamental nature of being in the cosmos, rather than merely seeking to describe the content of an objective morality. For Augustine and Aquinas, and their intellectual heirs, questions of morality were to be *derived from* the inquiry into reality and being, not separated from it.[1] Eventually, the attempt to ground the structure and content of law and government in broader credal positions was to collapse in the face of the fragmentation of Europe's religious inheritance in the sixteenth and seventeenth centuries. Thenceforth, newly envisioned nation-states confronted one another, not as sometimes quarrelsome brothers in an overarching spiritual community, but as sovereign enemies across the battlefield.

Within this new framework of thought and action, objective or cross-cultural morality could only appear as a set of standards to be achieved, rather than as a body of supra-historical norms awaiting

[1] See L Weinreb, *Natural Law and Justice* (Cambridge MA, Harvard University Press 1987) 15–39.

recognition and obedience. Treaties and Leagues proved inadequate instruments in the acknowledgement of absolutely binding standards. The moral disasters of the Second World War demonstrated at once the stark absence of a common source of values strong enough to restrain barbarism and appalling atrocity, but also the latent presence of senses and deep connections shared across the human species. Both: how can the idea of a universal moral order possibly be reconciled with such horrific violations; *and* how can such acts be treated as other than abhorrent to the inmost convictions of all decent beings?[2] It would be shallow to dismiss the various enactments and declarations of human rights that have taken place across the globe as simply illusory comforts or as instances of Western hegemony. But equally, one should not suppose that the existence of human rights instruments testifies to a normative order separate from and above that of the legal systems into which they have been integrated. Such an order, though it represents a flowering of thought and vision, is but a foreshadowing, and it is the more difficult and more lengthy task to achieve it.

The idea of 'human rights' harbours both conservative and anti-conservative elements. On the one hand, human rights are suggestive of commitments that are universal and fundamental in ambition. Implicit within the idea of essentially unchanging standards is a deep conservatism: the identification of forms and ideas that are non-negotiable and not to be included in the processes of political bargaining. However society might change or adapt itself to circumstances in other ways, certain standards are to be maintained as intrinsic to the very idea of civilized society. On the other hand, human rights are expressive of ideals that are more or less revisionary of established social institutions and practices. In so far as present forms of life fall short of perfectly instantiating the ideal (and it is perhaps in the nature of an ideal that this should be so), human rights *demand* social and political change.

The present volume of chapters relates the idea of human rights to the area of environmental law. Perhaps it is not strange that a taste for applying 'objective' standards to ever broader areas of law and life should develop at precisely the point where settled understandings and shared values are possibly disappearing from the world. In this chapter my purpose will be to consider how recent trends in environmental thinking should be situated within the context of legal concerns. My argument will be concerned with the general features of environmental

---

[2] Both responses are visibly and harrowingly present in M Perry, *The Idea of Human Rights: Four Enquiries* (New York, Oxford University Press 1998).

thought rather than with human rights, but on the whole it points to the conclusion that the association of environmental law with human rights promises overall to be a salutary one. Environmental law is conducted in the presence of similar oppositions of conservative and anti-conservative tendencies: the rhetoric of much environmental policy relates to 'conservation', but the intellectual standpoint of many environmental lawyers and campaigners is that of the reformer or the radical. One might be inclined to dismiss the importance of this apparent tension, for are not both tendencies very familiar aspects of experience? And what puzzle is there in the fact that our goals may lead us sometimes to conserve certain elements of established practices, and at other times to reform or revise aspects of those same practices? Nevertheless, I shall suggest that there *is* a question of some importance here, and it is perhaps the very familiarity of these dimensions of our experience that causes us to leave them unfathomed.

The ordinary concerns of environmental lawyers might at first seem somewhat removed from the issues I wish to raise. For it is not immediately obvious that *lawyers* need exhibit any particular awareness of the deeper political and philosophical debates which can inevitably be asked of any general body of legal rules. Environmental law, in perception at least, is still broadly a body of statutory law; and its administration involves the application of the measures and protections laid down in the statutory rules irrespective of the philosophical or pragmatic thinking which gave rise to their adoption.[3] Such thinking, it may be thought, is relevant to the political decision-making processes through which the law is created and amended, but must be separated from the adjudicative procedures in which the law is interpreted. For otherwise the adjudicative process will itself become politicized, and the courts transformed into general organs of judicial review through which previously decided political questions are opened up for revisitation.

One need not subscribe to 'legal positivism' in order to be troubled by this possibility.[4] Yet the law cannot be fully insulated from questions of the kind raised by broader philosophical debate. Famously, the common law operates not as a body of rules but as a body of reasons for decision; but this is no less the case with the application of statutory rules which,

[3] The perception of environmental law as largely a statutory affair may be misleading: see S Coyle and K Morrow, *The Philosophical Foundations of Environmental Law* (Oxford, Hart Publishing 2004) ch 4.

[4] Indeed, the depth of the worry has not been as fully appreciated as it might have been. For a prominent exception, see J Waldron, 'The Core of the Case Against Judicial Review' (2006) 115 *Yale LJ* 1346.

unlike the doctrinal principles of the common law, possess a fixed verbal form. For general rules must be interpreted in the process of being applied, so that the full context of a decision involves not only the written rule but also the specific significance that is given to it in the light of the case at hand. How widely or narrowly should the rule be construed? Would a strict application of it in certain circumstances frustrate its perceived purpose? General rules are inevitably rather blunt instruments, which stand in need of refinement: in addition to the rule, in other words, the application of law requires *judgment*. And in this way a body of jurisprudence will grow up around the fixed rules such that the conventions of *stare decisis* will become relevant to its application to subsequent cases. Thus, environmental law too exhibits the form of a body of reasons.

A systematic body of laws of any sophistication can be thought of as embodying a distinctive political vision, implicit in the structure of reasons that cluster around its provisions. But if the environmental lawyer's ordinary business of interpreting and applying rules cannot be carried on in isolation from thinking of a broader politico-philosophical kind, of precisely what character is that thinking? And in what manner should lawyers, in particular, take account of it? In answering these questions, I hope to shed some light, however obliquely, upon the themes of human rights and concern for the environment that define the present volume.

## 1. CONSERVATISM AND RADICALISM IN ENVIRONMENTAL THOUGHT

Tracing the general pathology of a pattern of thought is a notoriously difficult activity. 'Environmental thinking' is not defined by a single intellectual movement or collection of conclusions but by many partially overlapping and partly divergent ideas; and the development of environmental concerns is not exclusively or even primarily an intellectual product, but also a social one. Yet if the opening chapters of textbooks and collections of essays on Ecology, Environmental Philosophy and Environmental Law are in any way indicative, environmental thinking (among academic lawyers at least) is represented as a child of leftward-leaning political ideologies, and as requiring a 'radical' political stance. JB Callicott, for example, regards environmental thinking as part of the post-1960s' response to 'ivory tower', conservative analytical thought, a mode of applied ethics requiring an earnest attempt to highlight and resolve 'real-world problems'.[5] Because they concern the common environment, these problems necessitate the mobilization of the state's resources for the protection of a public interest, at the expense of private economic interests.[6] Federico Mayor, Director-General of UNESCO, in his foreword to a basic textbook on environmental law, points similarly to the 'far-reaching social implications' and 'radical changes' that organized responses to environmental problems call for;[7] whilst another textbook begins by noting that environmental thinking 'challenged the values associated with the idea of progress based on ever-increasing levels of economic growth …'.[8]

It is not difficult to connect environmental thinking with general issues of social justice, characteristic of left-political ideologies. RJ Lazarus (for example) locates the development of environmental law in the United States within the broader public health and urban justice movements of the 1970s, referring to it as the 'culmination of an era of protest'[9] against an unconstrained capitalist agenda. Notions of 'sustainable development', in so far as they relate to the protection of the interests of future generations, also exhibit clear involvement with concepts of social justice.[10] These associations with social justice themselves point to a basic trajectory in environmental thinking. On the one hand are those who suggest that the primary reason for instituting legal protection of the natural environment lies precisely in the resulting benefit to human communities. This view, in relating environmental protection to human needs and interests,

---

[5] JB Callicott, 'Environmental Philosophy is Environmental Activism: The Most Radical and Effective Kind' in D Marietta and L Embree (eds), *Environmental Philosophy and Environmental Ethics* (Lanham MD, Rowman & Littlefield 1995) 19–20.

[6] DM Driesen, *The Economic Dynamics of Environmental Law* (Boston MA, MIT Press 2003) 1–3.

[7] F Mayor, 'Foreword' in AC Kiss and D Shelton, *Manual of European Environmental Law* (Cambridge, Cambridge University Press 1997) xix.

[8] G Smith, *Deliberative Democracy and the Environment* (London, Routledge 2003) 2.

[9] RJ Lazarus, *The Making of Environmental Law* (Chicago, University of Chicago Press 2004) 51.

[10] See e.g., A Dobson, *Fairness and Futurity: Essays on Environmental Stability and Social Justice* (Oxford, Oxford University Press 1999). Arne Naess goes further, claiming that ecological principles concern not only pollution and resource depletion but 'deeper' principles of 'egalitarianism and classlessness': A Naess, 'The Shallow and the Deep' in A Drengson and Y Inoue (eds), *The Deep Ecology Movement: An Introductory Anthology* (Berkeley CA, North Atlantic Books 1995) 3.

constitutes a position that has come to be known as 'anthropocentrism'. Such a position is to be contrasted with one that is 'ecocentric' in seeking to protect or preserve aspects of the natural world independently of their relationship to human needs. The contrast between these two standpoints has been explained as follows:

> 'Anthropocentric', or human-centred, ethics argues that value in the natural world originates in the attitudes human beings take toward nature. Trees, waterfalls, and butterflies alike possess no moral standing beyond their capacity to serve human ends, or perhaps as repositories of values important to humans. In themselves, they have no meaning or ethically significant value … By contrast, a Deep Ecological or ecocentric view holds that plants, animals and sometimes even natural forms of land and water merit respect independently of their utility for or regard by human beings.[11]

There is a feeling among some environmental philosophers, and also some environmental lawyers, that the reasons which inform the creation and administration of environmental regulations must finally be informed by ecocentrism.[12] For such writers, there is a logical progression from the anthropocentric to an ecocentric view, which represents a deepening of the seriousness and commitment with which the natural environment is perceived as the object of moral duties and values.[13] It is no accident that

---

[11] G DiZerega, 'Empathy, Society, Nature and the Relational Self: Deep Ecology and Liberal Modernity' in RS Gottlieb (ed.), *The Ecological Community* (Abingdon, Routledge 1997) 57. It is not always possible to distinguish in practice between these two viewpoints: see e.g., Kenneth Maly's linguistic analysis of Aldo Leopold's *A Sand County Almanac* (1948): K Maly, 'A Sand County Almanac: Through Anthropogenic to Ecogenic Thinking' in BV Foltz and R Frodeman (eds), *Rethinking Nature: Essays in Environmental Philosophy* (Bloomington IN, Indiana University Press 2004) 289.

[12] See e.g., F Matthews, 'Deep Ecology' in D Jamieson (ed.), *A Companion to Environmental Philosophy* (London, Blackwell 2003) 218; JB Callicott, 'Intrinsic Value in Nature: A Metaethical Analysis' in *Beyond the Land Ethic: More Essays in Environmental Philosophy* (New York, State University of New York Press 1999) 239; PR Hay, *Main Currents in Western Environmental Thought* (Bloomington IN, Indiana University Press 2002); C Stone, *Should Trees Have Standing?* (3rd edn, Oxford, Oxford University Press 2010); A Gillespie, *International Environmental Law: Policy and Ethics* (Oxford, Oxford University Press 1997) ch 1.

[13] See in particular, R Routley, 'Human Chauvinism and Environmental Ethics' in DS Mannison, MA McRobbie and R Routley (eds), *Environmental Philosophy* (Canberra, ANU Social Sciences 1980) 96; W Grey, 'Anthropocentrism and Deep Ecology', 71 *Australasian Journal of Philosophy* (1993) 463; V Plumwood, *Feminism and the Mastery of Nature* (London, Routledge 1993).

proponents of this, more 'radical' view of the matter often draw inspiration from other 'radical' or 'postmodern' agendas generally associated with left-leaning political standpoints: either Marxist in origin, or ecofeminist.[14] For they share the same critical distrust of established hierarchies and willingness to 'deconstruct' traditional conceptions, and therefore embody the very antithesis of the conservative disposition. As John Gray observes:

> It is fair to say that, on the whole, conservative thought has been hostile to environmental concerns over the past decade or so in Britain, Europe and the United States. Especially in America, environmental concerns have been represented as anti-capitalist propaganda under another flag. In most Western countries, conservatives have accused environmentalists of misuse of science, of propagating an apocalyptic mentality, and of being enemies of the central institutions of modern civil society. Nor are these accusations always wide of the mark. Indeed, in considerable measure, they show conservatives endorsing the self-image of the Greens as inheritors of the radical protest movements of earlier times, and as making common cause with contemporary radical movements, such as feminism and anti-colonialism. In other words, both the Greens themselves and their conservative critics have been happy to share the assumption that socialism and environmental concern go together.[15]

Like Gray, although for different reasons, I wish to challenge this consensus. In order to challenge it, I must first dispose of the more radical 'ecocentric' versions of the position: not least because there are lawyers who have sometimes defended them.[16]

## 2. THE RADICAL VIEW

It might at first appear that 'ecocentric' views enjoy an affinity to the classical theories of natural law I spoke of earlier. For both involve the adoption of a cosmological perspective that seeks to understand a natural

---

[14] See especially, Plumwood (n 13); M Kheel, *Nature Ethics: An Ecofeminist Perspective* (Lanham MD, Rowman & Littlefield 2008); A Salleh, 'The Ecofeminism/Deep Ecology Debate: A Reply to Patriarchal Reason' (1992) 14 *Environmental Ethics* 195–216; I Balbus, *Marxism and Domination* (Princeton NJ, Princeton University Press 1983). Another instance of postmodern perspectives can be found in S Cataldi and W Hamrick (eds), *Merleau-Ponty and Environmental Philosophy* (Albany NY, State University of New York Press 2007).

[15] J Gray, *Beyond the New Right* (London, Routledge 1993) 124.

[16] See in particular, Stone (n 12).

order of which human life and concerns are but a part. But this appearance is deceptive. As with environmental thinking in general, 'ecocentric' thought embodies not one, but a number of more-or-less related positions. At its most basic, it advances the suggestion that human beings are not 'central to the cosmic drama':[17] that human life is not above nature but is part of nature. Whilst this proposition is not incompatible with certain natural law theories (those deriving from Aristotle rather than from Plato), those theories incorporated a view of nature as a created and teleologically ordered universe. Human life might exist in nature, but it is also involved in a normative order that is supernatural: our understanding of the world is thus ultimately to be drawn not from a perspective that is ecocentric, but one that is theocentric and reflective of that super-natural order.

More ambitious versions of the ecocentric view speak of ecocentrism as de-throning the Enlightenment conception of ethics, replacing an idea of morality as a purely human concern with one in which other dimensions of the natural world (animal life, tracts of land or water, and so on) possess value and form a source of moral obligations. For such writers, the 'challenges' of ecocentric thought precipitate a wider 'crisis of reason' as this has come down to us in the philosophical traditions of the West.[18] In its broadest interpretations, ecocentric thought urges upon the reader the necessity to abandon a purely 'human' view of the world, and to embrace a perspective that is distinctly 'other': a view widespread in postmodern philosophies of the environment, but also in the more mystical sections of Leopold's celebrated *Sand County Almanac*.[19]

Despite their flourishing among certain fashionable liberals, it is difficult to see how 'ecocentric' views, either in the extreme form of injunctions to 'think like a mountain' (and other, equally entertaining suggestions) or in the more general form of scepticism towards traditional philosophical ethics, can be taken seriously. From the earliest times, the Great philosophies of the Western tradition have sought to understand the nature of the human condition by contemplating its place within the wider cosmos. Aristotle's philosophy (for example) addresses the world of natural being. It seeks to understand the botanical properties of plants and the actions of animals as well as subjects such as ethics and

---

[17] M Humphrey, *Preservation Versus the People: Nature, Humanity and Political Philosophy* (Oxford, Oxford University Press 2002) 18 (quoting Warwick Fox).

[18] See in particular, V Plumwood, *Environmental Culture: The Ecological Crisis of Reason* (London, Routledge 2002).

[19] Leopold (n 11) 129–33.

'human sciences'. Indeed, the natural philosophy includes the study of Man, who is himself considered as a natural being. Much the same is true of Aquinas, Grotius, Hume and Kant: in each of these philosophies, the central question of human nature concerns the extent to which Man is a supernatural being (in having been created, unlike other animals, in the image of God); but in none of them is the human being understood wholly outwith the context of the created order of nature in which he appears. Even in the work of St Augustine, for some the most Platonistic of Christian thinkers, mankind is represented as having a complex and divided nature: a being both of spirit and intellect, on the one hand, and of flesh and worldly passion on the other.[20] Though it no doubt appears here and there, the motif of Man as being 'above nature' in no measure identifies a general seam of thought running through this intellectual tradition. What appears with more consistency, though again in conjunction with its denial, is the view of the human being as, in certain respects, *apart from* nature (and perhaps nowhere more so than in Augustine and in Protestant writers in the Lutheran tradition during the early-modern period.) This 'apartness' carries no intrinsic sense of superiority (or 'aboveness'), but is in my view central to ethical understanding.

The relevant lesson to be drawn from Aristotle might be stated as follows. As with his general philosophy, Aristotle's political philosophy also begins with nature. The state, Aristotle says, is a creation of nature, because 'man is by nature a political animal'.[21] This political nature separates human beings from other animals: men are not simply *social* animals, but are capable of exhibiting choice about the form of their social relationships. In the first place, the survival of the human race obviously depends upon human beings forming unions and becoming what he calls 'companions of the manger' and later 'companions of the cupboard'. Families, therefore, are natural units in the sense that they are necessary for the bare requirements of life. But the nature of a thing is defined not by what is merely necessary to its existence, but rather 'what each thing is when fully developed'.[22] And it is the fact that human beings live a life in which they are perpetually confronted by choices concerning their development that marks them as proper subjects of ethics. The structure of ethics (pursuit of the good life) is thus naturalistic in being grounded in the human *telos*; but it is at the same time proper only to humans (for only they, unlike other animals, are capable of

[20] See Augustine, *City of God* (Harmondsworth, Penguin 1984) ch XIX (and elsewhere).

[21] Aristotle, *Politics* I.2.

[22] Ibid.

denying their *telos*). This does not prevent other forms of animal or plant life from becoming the object of moral concern; but the genesis of that concern lies not in the possession by the object of certain qualities or values (still less of *rights*),[23] but in the possibilities present within human action when directed towards such objects.

One need not embrace teleology in order to grasp the basic distinction that is at issue. It is the presence of a political nature that marks humans as moral beings, and the status of human actions as genuine potentialities or breaking-off points in history that is the ground of norms and values. The separation of Mankind from the rest of nature is thus as important as its situation within a wider order of nature.

In what sense does the human being 'stand apart' from nature? Aquinas thought the answer lies not in the spiritual composition of the human being but rather in the manner of its language. Aquinas regards all living things, animal or human, as having a soul (or 'anima'). The possession of an anima is the possession of the ability to move oneself, to be self-moving. Human beings and lions share this quality properly and not merely analogously. To be animated, or alive, is to constitute an organic structure: that is, a unit in which all parts are connected organically to the whole. Where this is the case, the various parts (of the lion or the man) are characterized by reference to the whole, and not vice versa. So, if one were to strike the face of the man, one is striking the man; if one strokes the lion's fur, one is stroking the lion. Now, we might think of the world as impacting upon the lion in numerous different ways. It senses movement through its whiskers, sound through its ears, pain through collisions of its outer surfaces with external objects, and so on. If the lion constitutes an organic whole, then the different parts of the lion (its eyes, ears, skin, etc.) are its organs. And we can think of these organs as rendering the world meaningful to the lion through the manner in which they are affected. The world, that is, becomes significant for it by the way in which it is affected: by pain, hunger, thirst, tiredness, fear, pleasure, desire (for what is seen), etc. It is on the basis of the lion's ability to undergo these various sensations that we may speak of it as being aware. Lions act wilfully and deliberately: they decide to chase the prey, or abort the chase; they do not wander or sleep aimlessly, but with purpose; and in their relations with others of their kind, they fight, submit, flee, nurture, protect or harm, not as pure automata, but as living things with a will of their own. At the same time, however, there is no

[23] See Stone (n 12) for the classic argument (that 'trees have standing'); but it has been oft repeated and emulated in writing on 'deep ecology'.

morality in the lion's behaviour: it is not evil when it attacks its prey (or other lions), nor does it do wrong in the killing of newborn cubs to make the mother lioness go back on heat.[24]

The lion perceives and interprets the world on the basis only of sensory meaning. Because of this, its range of responses to its situation is limited. As the human being is born into society, and has no choice about that, so the lion is born into a pride, and has no choice but so to be. The structure and functioning of the pride, however, is determined by the genetic background of the lion as a species. That is, the mode of interaction between lions within the pride, and the pride's relations with the rest of the world, are a function of instinctive responses shaped over millennia of evolution. A lion does not *learn* what it must do to survive an attack on its body, or what arrangements it must make to nurture its young. Very roughly speaking, the pride's survival is determined by evolving those responses which 'work', as opposed to those which tend towards the extinction of the species. Any changes in the structure and characteristic activities of the pride take place therefore only over the very long term. (We can still think of these changes as having their origin in the will, although they are not in any sense purposive.) Like the lion, human persons also have a range of options open to them in the light of the situation into which they are born. And this situation too is essentially social. But the difference is not merely that human beings have a much greater *range* of responses open to them in their mode of belonging; it is that the responses are purposive rather than instinctual. To understand this, we must contemplate the nature of the human being as a *linguistic* animal.[25]

Lions interpret the world on the basis of sensory meaning. Human beings, as distinctively *rational* animals, are additionally able to interpret the world through linguistic structures of meaning that they themselves have created. To perceive the world through language is to regard it from the perspective of the possibilities it offers. It is the ability to interpret the world as falling under a great many descriptions, and thus as presenting choices as to one's mode of engagement. The man standing over there is *both* potentially a threat *and* a potential friend, *and* a useful pawn in my larger plans, and a whole host of other things. When I act, I do so not on the basis of a programmed genetic template forged over millennia, but in virtue of my description, that is, in virtue of the world as it appears to be

---

[24] One is reminded of CS Lewis's remark in *The Lion, the Witch and the Wardrobe* that Aslan (who might here stand for nature itself) is 'not a tame lion'.

[25] For a useful, and longer, discussion of some of these themes, see H McCabe, *On Aquinas* (London, Continuum Books 2008) 33ff.

relevant to me in my thinking of it. Language changes everything: we must not think of human beings as perceiving the world first instinctually *and then* linguistically (as if we first have sensations and then give a name or linguistic meaning to them). Rather, the character, or meaning, of our sensations is changed forever, once we have mastered language. Though it is in some ways determined by its belonging to a species, the lion's sensory interpretation of the world around it is a private affair. What it sees and feels is uniquely known to it and to nothing else. But language is a public phenomenon, the words out of which we construct the world (as it is known to us) deriving their meaning not from my private association of individual words with my particular inner-sensations (or mental images), but from their usage in the language as a whole. They are words *of English*, not words *of mine*.[26] (We might recall Hobbes's remarks on the state of nature here. The loss or absence of language was one of the things which characterizes the natural condition of mankind as a war of each against all; a state in which, as Hobbes says, man becomes a wolf to his fellow man.)[27] It is this linguistic ability that defines human beings as rational animals.

For human beings therefore, there is not 'the world' and then an 'interpretation' of it; not 'experience' and then its intellectualization. Rather there is no separation between our experience of the world and the meaning we give to it: the facts are always a particular interpretation of the facts.[28] Because experience everywhere involves judgement, and cannot be finally separated from it, there exists no platform from which we are able to abandon a 'human' view of the world and adopt a perspective from which human concerns, interests and structures of meaning and understanding are not central. Language is in one sense a response to the way in which the world impinges upon the human being; its meanings and concepts presuppose this connection and cannot be

---

26 Many philosophers (the so-called 'analytical Thomists') have therefore stressed the relationship between Aquinas and the twentieth-century analytical philosopher Ludwig Wittgenstein.

27 T Hobbes, *Leviathan*, R Tuck (ed.) (Cambridge, Cambridge University Press 1996) ch 13. Strikingly, both morality and society are said by Hobbes to be absent from this natural condition, just as Aquinas regards morality and society as being particular to human beings as linguistic animals.

28 For an excellent discussion see M Oakeshott, *Experience and its Modes* (Cambridge, Cambridge University Press 1933) ch 2: 'It is not wrong, of course, to … distinguish (for example) sensation, reflection, volition, feeling and intuition; the error lies in supposing that in so doing we are considering activities which are different in principle and can be separated from one another finally and absolutely' (10–11).

severed from it. I rehearse these points in order to counter the post-modern suggestion, adopted by the more radical ecologists, that a successful environmental ethics requires the adoption of the standpoint of 'the other'. The human capacity for sympathy and imagination should not be mistaken for this possibility. When the Lord reminds Adam, 'Dust thou art, and unto Dust thou shalt return',[29] it is not as an imaginative sensation of an altered state that the image has meaning for us, but only as a reminder of the finitude and briefness of the present set of experiences and sensations which constitute our life. In the same way, we possess no conception of value that makes sense other than in terms of these experiences and sensations.

What, then, is the appropriate meaning to be given to the idea that the natural environment is of 'intrinsic' rather than 'instrumental' value? At one level, it is the sense that an object is valued purely in its givenness rather than for any reasons why it should be useful for *particular* purposes. The status of the planet as a 'home' for human beings captures something of this sense: it is valued through our belonging to it, prior to and independently of any specific uses we make of it in the advancement of concrete goals. (Of course, this sense of belonging is in part a consequence of its *general* suitability for human life.) One can imagine such ideas as a ground of ethical duties of preservation (or cultivation) which do not have as their object any particular utilitarian end in view. And it is, in a certain sense, based on a view which values the natural environment 'for its own sake' (much as one might pursue knowledge for its own sake, as something that is simply good in the form in which it is given).[30] But it is *not*, at the same time, a conception of a value that is fundamentally separated from the world as it is known to and valued by, human beings.

The hopes of the 'radical' ecologists therefore rest upon a mistake. For the appropriate sense in which one might point to a dimension of value in the natural environment that is unknown or wholly irrelevant to human existence is not to be found in any 'postmodern' romanticism, but in transcendent religion. Radical 'ecocentric' thought began from the assumption that environmental concerns must be derived from an ethical standpoint that requires us to give up a distinctly human perspective. Such a standpoint is not available. What in fact is required is a standpoint from which ethics does not terminate in human concerns, but is properly

---

29 Genesis 3.19.

30 See e.g., J Finnis, *Natural Law and Natural Rights* (Oxford, Clarendon Press 1980) ch 3.

transcendent of them. Ethics is inherently rooted in distinctions, both as to what is valued and under what conditions it is valued: we value certain things *and not* others. That which possesses 'absolute value' is not valuable *without* condition, but inclusive of *all* conditions. Ethical values, in other words, belong to a scale of values; and if this scale is not related to conditions of human interest, then it must draw its being from an interpretation of the world as part of a cosmos that is intelligible in other terms. For if the cosmos is thought to be, save in human terms, orderless and purposeless, then in what respect can it be said to possess any but instrumental value? The idea of a 'rational order of nature' is articulated most clearly in the idea of a Created cosmos, as it appears (for example) in Christian theology. But it is also present within Confucian and Daoist traditions, for example, in that the scholar who seeks 'oneness with nature' is seeking to locate himself within an ordered and balanced system; and it is the balance itself which is of ultimate value.[31]

But if this is the only way in which sense can be made of the environment as possessing 'intrinsic value', we may nevertheless ask how such ethical concerns are to be accommodated within the conceptual structures of law and politics?

One cannot simply adjust one's radicalism to incorporate a religious basis (in natural law) in place of a non-human 'ecocentric' basis. Such a move will render the radical's position coherent where before it was not, but it will *not* assist the reception of the appropriate range of ethical concerns into legal and political thought, for it is not simply the ecocentrism, but radicalism itself that is ultimately problematic.

The essence of radicalism is the insistence that an established set of values and practices be forced to yield to an alternative scheme of values that is foreign to it. Such cosmological views of nature are not, however, capable of straightforward reception into the political and legal thought of the West. For modern political culture, as it developed in post-Reformation Europe, is founded upon a deliberate separation of the realm of law and politics from that of religious credal belief: a separation intrinsic to the Protestant worldview out of which the modern state

---

[31] See O Koichi, *Nature and the Views of Nature in Chinese Literature* (Shanghai, Shanghai Ancient Books 1989). On the Daoist view, see the *Dao De Jing* (various eds) ch 25: 'There was something formless yet complete / That existed before heaven and earth / Without sound, without substance / Dependent on nothing, unchanging / All pervading, unfailing. / One may think of it as the mother of all things under heaven. / Its true name we do not know; / "way" is the by-name that we give it'.

developed,[32] and enshrined in Article 10 of the Charter of Fundamental Rights of the European Union (UK HRA Art 9). A patient and conservative mind may discover, in established legal concepts of property, right, obligation and so on, the faint echoes of these religious concerns, and may even undertake the fragile restoration of the fading system of values which informed their development.[33] But the radical, who perceives the relationship between such values and the political system as one of opposition, in which the latter must ultimately yield to the former, actively obstructs the possibility of a restorative project. A 'radical' standpoint on politics is one that is mistaken in supposing that the activity of politics consists in effecting the transformation of social arrangements and institutions in line with abstract ethical ideals. The idea that the political system exists in order to give effect to an ethical viewpoint is not a useful or accurate point of departure for reflection, however. The nature of political action, in all but totalitarian regimes, is such that it demands a mandate for power that is rooted in a significant body of sectional interests.[34] In societies of a relatively open character, bodies of such interests represent, not homogenous opinion, but a complex series of distinct opinions which converge and overlap at various points, whilst remaining strongly divergent at others.

The idea that a distinctive moral position can be inferred from such a mass of opinion may count as one of the central myths of modern political theory. Such political theories by-and-large fail to take account of what Raymond Geuss has called the 'differential structure' of political thought and action.[35] If we think about 'applied ethics', we tend to conceive of an idealized discussion (of the kind envisaged by Habermas, Rawls or Dworkin for instance) in which a topic is to be explored fully, with no time-limit in mind, and in which earlier topics can be revisited without restriction in the light of changes of view or order of consideration. But politics cannot be 'applied ethics' in this sense: for concrete agents operating in the real world of scarce resources and shifting bases (and mandates) of power, the decision to pursue one thing means the

---

[32] For a full discussion see S Coyle, *From Positivism to Idealism* (Aldershot, Ashgate 2007), and the excellent discussion in I Hunter, *Rival Enlightenments* (Cambridge, Cambridge University Press 2001).

[33] See Coyle and Morrow (n 3), and sections 3 and 4 below.

[34] Totalitarian regimes may themselves demand a degree of sectional support from the population, though perhaps in lesser degree: see Coyle (n 32) ch 4.

[35] R Geuss, *Philosophy and Real Politics* (Princeton NJ, Princeton University Press 2008) 30–31.

failure to choose and pursue another.[36] Unlike the ideal discussion, in politics '[o]nce I have done something, I have changed the situation, sometimes for good'.[37] Politics has a historical trajectory which prevents the emergence of genuinely freestanding, abstract possibilities. It is an ethical activity (i.e., a subject for moral philosophy) not because it consists in applying abstract theories to real-world conditions, but because all of its activities have moral consequences, and its purposes invite moral analysis.

For all of these reasons, a radical viewpoint is not a useful starting-point for thinking about environmental regulation. I have criticized 'ecocentric' views as being misguided and mistaken about the nature both of human cognition and of morality. And I have argued that, in the only form in which sense can be given to it, the suggestion that an environmental ethic propounded in separation from human concerns should guide the interpretation of law and politics, is not assisted by the deepest assumptions of a 'radical' standpoint.

## 3. CONSERVATISM V. FRAGMENTATION

I suggested earlier that doctrinal legal argument cannot ultimately be isolated from broader ethical concerns. What form must such concerns take, if they are not those of the radical? We might begin by reflecting upon the relationship between legal order and other aspects of society. Law is but one facet of social order: government, collective institutions such as the market, and patterns of family life all contribute to the overall structure of society. Not one of these institutions has final priority over the others. Governments shape the law, both through the legislative mechanism and (sometimes) the appointment of judges; but law also limits the actions of government through the constitution and the rule of law. The market is controlled by both law and government; but it has an independent existence which is also beyond the control of either, requiring that it be responded to and nurtured as well as curtailed and interfered with. Ordinary family and social life are subject to law and government, but both law and government must in some measure offer protection to, and not suppress or revolutionize, settled patterns of conduct.[38] These are familiar facts. But they point to an understanding

---

36 Ibid.

37 Ibid 31.

38 See Lon Fuller's classic discussion of the rule of law: *The Morality of Law* (New Haven CT, Yale University Press 1964) ch 2.

of the nature of law that is less familiar: one that is nevertheless important in the context of environmental ambitions.

A little understood aspect of law is that it functions both as an instrument of conservatism *and* as an agent of change. The relationship of law to governance implies the coexistence within law of both these forces; for we think of one purpose of government as being to advance society in the direction of certain political goals, using law as its instrument, but law also forms a layer of control on the exercise and extent of governmental powers to bring about changes to the existing conditions. This dual aspect of law might be viewed as a property of all law, or as being intrinsic to the very idea of law, so that it is impossible to distinguish areas or types of law that are progressive from those which seek to rein in the forces of progress through the defence of the status quo.

The association of environmental thinking with radical or anti-capitalist agendas has lent to environmental law the appearance of a progressive body of rules, seeking to challenge orthodox assumptions about individual and governmental rights. In this context, the dominance of human (largely economic) interests over the natural environment has been perceived as one more hierarchy to be confronted or reversed; a pattern of thinking that has allowed for the emergence of movements such as 'eco-feminism'. Associations of this kind naturally portray environmental measures as innovations standing in fundamental opposition to 'established interests', such as property rights and the operations of the market. The relevant image is one in which the law reflects the conservative agenda of 'the establishment', in protecting its entrenched hierarchies, so that new and innovative law, driven by a different, anti-establishment framework of priorities, is required in order to dismantle them. This misreading of the law's conservative nature has many sources, of which perhaps the most significant is too close an identification of legal order with the prevailing pattern of social order and custom. As we have seen, the law must inevitably offer a level of protection to settled practice and interests (lest society dissolve in permanent flux); but it is also distinct from and frequently in opposition to such interests (lest society dissolve through stagnation). An image of property rights and free-market operations as intrinsically unbounded manifestations of the unhindered spirit of capitalism until interfered with by socialist legislation, is not one that features in legal thought (if it truly features anywhere).[39]

---

[39] On Thatcherism's paradox (wherein the policy of laissez faire demands more direct and all-encompassing governmental intrusions into society and the

The legal order is not an instrument of capital: it both supports capitalism, and contains it within certain limits. Despite the tendency of modern capitalists to regard them as such, the institutions of property and the market do not exist purely as wealth-producing mechanisms, but as the social forms of a living community. They are not detached and free-floating, but grounded in a shared form of life. We are used to thinking of free-market institutions as bringing about the desolation of communities, due to the prioritization given to open competition over notions of public duty or service to the common good that has been a feature of governmental policy (on both the left and the right) for a generation. Competition is naturally hostile to institutions of 'civil society' (such as the family, the professions and the Church) in so far as they are constituted on other than market considerations: where such considerations intrude, they will seem to embody restrictive practices, precisely because they impose standards (of both behaviour and membership) that are not based on price. Consequently, governments since 1979 have presided over the gradual destruction and removal of those intermediate institutions (professional associations, local government competencies, labour movements, to name but a few) whose function it was to mediate between the individual and central government. A society recently denuded of such institutions will struggle to find, amongst its professional classes, any who have a direct interest in the maintenance and renewal of its common environment.[40] Attention to aspects of the common environment, in such circumstances, will become the province of *issues* rather than of *interests*. Thus, environmental law will appear on the scene as a radical and novel alteration to patterns of entitlement and practice that are seemingly built into the fabric of society itself.

This (rather unhistorical) view of the matter is apt to disguise the law's role in protecting and renewing the form of social life over which it presides. The legal instruments that regulate and promote market institutions also serve to anchor those institutions to the values and interests of civil society (where they properly belong). There is a doctrine of freedom of contract precisely *because* there is also regulation of contract,

---

market), see N Boyle, *Who Are We Now? Christian Humanism and the Global Market from Hegel to Heaney* (Notre Dame IN, University of Notre Dame Press, 1998) ch 1.

[40] Gray presents this as a key problem of centrally planned economies in the old communist states, due to the absence of the price-mechanism; but the same is true in capitalist societies in which competition has replaced or gained priority over all other considerations, so that there is *nothing but* the price-mechanism. See Gray (n 15) 130–33.

and property rights exist because there are limits to such rights, which prevent their exercise from becoming anti-social. In *The Philosophical Foundations of Environmental Law*, Karen Morrow and I attempted to draw attention to those aspects of environmental law that have their roots in the law of property, and of tort. The nineteenth-century law of tort presents a surprisingly sophisticated attempt on the part of the courts to recognize the nature and limits of property rights in the light of their manifold social impacts. A close study of that case-law reveals the emergence of environmental concerns growing out of rapid social changes, particularly relating to new forms of land user which arose from industrialization.[41] Upon reading the cases, the impression gained is not one in which the courts are seeking to uphold an established hierarchy against narrowly-formed limiting principles (private property v. environmental nuisances), but one in which a genuine effort is being made to reconcile a number of more-or-less conflicting interests and users, none of which has ultimate priority.[42]

A society naturally oscillates between periods of rapid change, and longer periods of stability and settled practice. It is likely to be in the times of upheaval and uncertainty that the conservative dimension of law is uppermost, for it is here that some mechanism is required which can assimilate new directions and concepts to existing forms of practice, and which can also adapt the familiar ways of thinking and acting to the new and changed realities. The industrial revolution exemplifies one such period; the 'social revolution' under Thatcherism and its successors is another. If this dimension of law has seemed less pronounced in modern times, it is only because law itself was employed as the principal instrument of change, a situation more or less unprecedented on such a scale in British society.[43] In such circumstances, the dual nature of law (as both instrument of conservatism and agent of change) is more pronounced, but also less clearly perceived. For we tend to conceive of statute law as expressive of aspects of governance that are forward-looking and progressive, whereas common law seems more conservative, slow to change, preferring to explore and adapt old ideas rather than abandon them in favour of new ones. But when rapid social change

---

[41] See Coyle and Morrow (n 3) esp. ch 4. There is no need to repeat the lengthy analysis here.

[42] See in particular *St Helen's Smelting Co v Tipping* [1865] 11 *HLC* 642, and the analysis in Coyle and Morrow (n 3) 114–15.

[43] One might compare this situation to that of the industrial revolution (for example), where the principal agent of change was technology, and entrepreneurship.

undermines familiar structures and contexts, lawyers may find themselves without the luxury of time to develop a measured reconciliation of all the conflicting implications of practice, so that legislation itself supplies the conservative response that is nearest to hand. When such developments occur, the mediating effect of an emergent legislative framework might easily be mistaken for a 'radical' revision of existing 'settled' doctrines. Such, one suspects, is the case with much environmental legislation.

The character of environmental legislation has much in common with a conservative disposition. A conservative, as Michael Oakeshott famously said, is one temperamentally disposed to value what presently exists over things which yet may be. There is nothing historically special about the present condition of things; it is not a fortress into which one must retreat if society is to survive, but simply the latest in a line of unrepeatable contingencies. Yet it is through such successive states of contingency (in unbroken line) that the identity of a community, and its inhabitants, is grounded. The conservative does not seek to *prevent* change, therefore (and thereby court stagnation), but rather to accommodate change, partly by preserving familiar features of present experience that are not threatened with extinction, and partly by assimilating what is new to those established features. The conservative recognizes that:

> [i]nnovating is always an equivocal enterprise, in which gain and loss (even excluding the loss of familiarity) are so closely interwoven that it is exceedingly difficult to forecast the final upshot: there is no such thing as an unqualified improvement. For, innovating is an activity which generates not only the 'improvement' sought, but a new and complex situation of which this is only one of the components. The total change is always more extensive than the change designed; and the whole of what is entailed can neither be foreseen nor circumscribed.[44]

As John Gray has pointed out, this attitude fosters an inter-generational perspective on politics that has particular resonance for environmental thinking.[45] It is here that the deepest affinities of environmental and conservative thought are revealed. For both are hostile to the central myth of modern politics: that of perpetual economic growth. It has been the idea of unlimited growth that has, in modern times, stimulated the detachment and cultivation of market institutions in separation from the communities that ultimately give those institutions meaning. The risks inherent in so detaching the market from its social context (the

---

[44] M Oakeshott, 'On Being Conservative' in *Rationality in Politics* (revised edn, Indianapolis, Liberty Fund 1991) 411.

[45] Gray (n 15) 136.

creation of globalized economic interests to which the individual 'consumer' stands in unmediated relation)[46] are only now beginning to become well understood. Its transformation of the individual from a being with particular purposes and interests, and a definable place in life, into an endless consumer of goods, is perhaps the most damaging of these because it 'encourages us to view our lives, not under the aspect of eternity, but as moments in a universal process of betterment'.[47] If the Thatcherite and Blairite policies of recent decades have encouraged these tendencies in the sphere of political economics, they have featured no less heavily in the intellectual writings of the liberal left. Agendas for social justice are similarly remarkable for their presentation of Western liberal societies as stages on the way to a more developed existence that lies further down the road: Raymond Geuss has drawn attention to the propensity for jurisprudential theories such as that of John Rawls to function as 'compensatory fantasies' wherein the shortcomings and injustices of the present are transformed into a theory of the ideal society that is essentially similar in its broadest aspects to the society of the moment.[48] Attainment of the ideal is then a matter of refining particular aspects of social institutions and mechanisms of distribution that stand at 'the end of history'.[49] The effect of such theories is to represent the society and experiences of the moment as impoverished or artificially limiting: there is a mode of perfection historically available to us, but in the form of a *postponed* good.[50] One key assumption, common to both environmental and conservative thinking, however, is that '[t]he characteristic that best distinguishes flourishing ecosystems is never growth, but rather stability …'.[51]

One might think of environmental thought as one aspect of a conservative viewpoint in this regard: it expresses with more than usual clarity and precision the risks involved in extensions to our characteristic patterns of activity that have developed over the last century, and which

---

[46] Indeed, perhaps the more damaging but less remarked-upon relationship is that between global corporations and the individual *producer*.

[47] Gray (n 15) 139.

[48] See R Geuss, *Outside Ethics* (Princeton NJ, Princeton University Press, 2005) chs 1–3, and J Rawls, *A Theory of Justice* (revised edn, Cambridge MA, Harvard University Press 1999) ch 1.

[49] F Fukuyama, *The End of History and the Last Man* (Harmondsworth, Penguin 1993).

[50] I have discussed the eschatological and political implications of this view in 'The Limits of Legal Ideologies' (2009) 60 *Northern Ireland Legal Quarterly* 471–86.

[51] Gray (n 15) 138.

continue to develop in ways that are not always transparent to us. It also seeks methods of assimilating innovations that threaten the survival or the renewal of the common environment. It worries about the long-term losses that such changes threaten to bring. None of these are 'radical' concerns; they appear so only because of an ancient paradox whereby the conservative, faced with rapid social change, finds himself displaced from the main current of political argument (with which he is naturally in sympathy) and doomed to adopt the character of the rebel.

The lessons for the doctrinal lawyer are not easy to draw from these considerations. An increasing tendency for core areas of legal practice (contract law, property, even tort and restitution) to be reinterpreted as aspects of, or subject to overarching principles of, commercial law, is explicable in terms of the politics of the day, but unlikely to leave room for the interpretation of environmental concerns as intrinsic to property rights or to practices of bargaining. A certain fragmentation of legal experience is likely to result, so that environmental (and other social justice) concerns seem to require formulation in detachment from bodies of rights with which they seem to be in tension. Such a development may halt the separation of the market from social concerns for a day, but it will not bring about the stability that is required for a long-term reconciliation. Rather, successive battles are likely to be fought over the extent of regulatory 'intrusions' into free-market practices and ownership rights; now resulting in strengthened regulations, now their erosion. Perhaps it is the lot of the doctrinal lawyer simply to preside over these battles, whist remaining powerless to address the assumed division that has led to them. If that is so, the most that can be suggested is that the connection of modern environmental principles with the deepest aspects of property law and tort must be continually asserted. The price for not doing so is the gradual erosion of inherited concepts, and the transformation of the legal order into a system of adventitious adjudication between competing instrumentalities.[52]

## 4. THOUGHTS ON HUMAN RIGHTS, LAW AND THE ENVIRONMENT

The fragmentation I have described is a consequence of those anti-conservative forces within politics which, rather than seeking the deeper

---

[52] It represents, in other words, the loss of what Nigel Simmonds has called 'the depth of the ideal': see NE Simmonds, *Law as a Moral Idea* (Oxford, Oxford University Press 2007) 145–49.

integration of a form of social life, go off in search of an ideal re-ordering of society; a re-ordering which is nevertheless understood in so many different and non-compossible ways. I mentioned at the beginning that the association of human rights doctrines with environmental law may ultimately be salutary rather than the reverse. What therefore is the likely effect of human rights law upon these developments, as they affect environmental regulation? Though human rights instruments are themselves subject to anti-conservative as well as conservative forces, it may be that the doctrine of human rights allows for the expression of conservative ideas in a more direct way than is possible within the current circumstances of party politics. A tentative, optimistic and inevitably highly inconclusive description of this possibility might run as follows.

The idea of immanent order which characterized Greek and early Christian philosophy long ago disappeared from the world, as it is understood politically. In its place arose a conception of 'objective' moral order that stands apart from the earthly realities over which it rules in judgement. The inspiration for the new framework of thought was originally Protestant: on the one hand was the sinful world of flesh and passion, and on the other a set of perfect standards to rule over it and reveal its sinfulness and imperfection; it seemed axiomatic that the prosaic, corrupted world could offer no insight into the nature of the good, and that the ruling standards could emerge only from a suitably attuned conscience.[53] With the declining influence of religious beliefs in matters of morality and politics, this disappearance of meaning from the world, and its location in the individual will or conscience, encouraged forms of subjectivism in ethics. The understanding of society appropriate to this changed picture is not the classical one of a form of association that exists to realize a complex and multi-layered idea of the good, but that of a free-for-all in which each person is a separate economic unit with her own preferences and priorities, and in which politics must achieve a kind of averaging between the mutually opposing interests. A conception of society along these changed lines makes fragmentation almost inevitable: its preoccupation with subjectivity tends to absolutize interests that were better conceived as more limited aspects of a single good that is complex without being internally contradictory.

The doctrine of human rights is (perhaps) the first tentative step, in a world shorn of religious ideas, on the road to dispelling the image of the person as an *ens completum*, forming a quite separate locus of interests,

---

[53] I have dealt extensively with these themes elsewhere: see Coyle (n 32).

ideas and priorities. For it suggests that there are certain common conditions that are required if such beings are to flourish in this world, and that such flourishing can therefore be achieved or promoted only as a common endeavour, and not as a result of the several uncoordinated efforts of disparate individuals. A clean and habitable environment is certainly one of these commonalities. But then the identification of such commonalities cannot be thought to constitute a systematic framework of principles that are *imposed* upon otherwise absolute interests (such as property interests) that might arise from the minds of radically distinct and therefore equal individuals. Rather, such a framework must be considered as the background of understanding against which the individual interests derive their meaning and significance: such interests must appear, not as radical inventions, but nuances in relation to the interpretation of common themes.[54]

It is (or was) the function of law to articulate and give precise definition to this shared background of understandings. Though it is possible to specialize in property law, contract, or tort, these conventionally distinct areas are not unconnected; nor are the boundaries between them 'natural' or set in stone, but capable of being understood in various ways:[55] rules on what can be contracted depend upon doctrines of ownership and transfer; the exercise of property rights is limited by tort, and by the criminal law; and so forth. Conventional boundaries between areas of law arise out of the necessity of interpreting the mass of rules and decisions relating to tort (say) as possessing internal coherence and systematicity; but they are not meant to deny the coherence and continuity that necessarily exists *between* the different areas of law. For these subject areas are not self-standing bodies of arbitrarily commanded rules, but aspects of a common form of life that is ultimately integrated and continuous.

Modern understandings of the liberal democratic state threaten this sense of integration, for they are apt to mistake it for a mere averaging of the competing interests of 'consumers' whose potentially destructive interference with one another when pursuing disparate goals stands in need of regulation. The liberal society is based on a common form of life only in a reduced sense: as a series of bodies in proximity rather than of

---

[54] The dependency of ideas of property on socially constructed dimensions of meaning and practice was a firmly embedded aspect of the writings of both Hobbes and Grotius, for example.

[55] See NE Simmonds, 'The Changing Face of Private Law: Doctrinal Categories and the Regulatory State' (1982) 2 *Legal Studies* 257–68.

minds in relation.[56] In place of a sustained pursuit of a given bearing, it is subject (particularly in its democratic form) to rival and ever-changing constituencies. Much has been said of the corrosive effects of 'individualism' in the modern age. In this essay I have traced some of its consequences for the fragmentation and absolutizing of interests within law. The doctrine of human rights promises, though no more than partially, to return us to something closer to the pre-modern conception of the person: as a being with a unique and individual soul, for which he alone is responsible, but which is embodied in a fleshly home whose essential characteristics and problems, and whose ultimate well-being, are inexorably linked with others of his kind. Human rights emphasize the collective context in which human life runs its course. In so doing, they may perhaps produce a more integrated understanding of law; and of rights that are ultimately tempered by the needs of a common environment.

---

[56] I borrow the expression from M Oakeshott, *Rationalism in Politics* (revised edn, Indiana, Liberty Fund 1991).

# PART 2

# Reconfiguring the legal

# 5. Human rights and environment protection in India: the judicial journey from public interest litigation to the National Green Tribunal

**Gitanjali N Gill***

## 1. INTRODUCTION

The relationship between human rights and environmental protection is not linear: it has not progressed harmoniously; sometimes its nature and scope has been vigorously debated; sometimes a collision or conflict was inevitable; and at other times its implementation was – and remains – questioned.[1]

* I recognise and thank the British Academy/Leverhulme Small Research Grants for supporting this research.

[1] A Kiss and D Shelton, *International Environmental Law* (Nairobi, UNEP 2005) 369; A Boyle, 'The Role of International Human Rights Law in the Protection of the Environment' in A Boyle and MR Anderson (eds), *Human Rights Approaches to Environmental Protection* (Oxford, Oxford University Press 1996); M Thorme, 'Establishing Environment as a Human Right' (1991) 19/2 *Denver Journal of International Law and Policy* 301–42; CJ Dias, 'Human Rights, Development, and Environment' in Y Danieli, E Stamatopoulou, and CJ Dias (eds), *The Universal Declaration of Human Rights: Fifty Years and Beyond* (Amityville NY, Baywood Publishing 1999) 395–402, 400; L Collins, 'Are We There Yet? The Right to Environment in International and European Law' (2007) 3/2 *McGill International Journal of Sustainable Development Law and Policy* 119–53; K Fatma-Zohra, 'Human Rights, Environment and Development' in S Lin (ed.), *Sustainable Development* (Nairobi, UNEP 1995); A Boyle, 'Human Rights and the Environment: Where Next?' (2012) 23 *EJIL* 613–42; DK Anton and D Shelton, *Environmental Protection and Human Rights* (Cambridge, Cambridge University Press 2011).

This chapter focuses on how the Indian judiciary has applied a human rights lens to analyse and address environmental protection. Two human rights, the right to a healthy environment and the right to development are examined in this context. Recognising and promoting the intertwined relationship between human rights and environmental protection is crucial. The developments presented in this chapter chart the approach adopted by the Supreme Court of India. The court has locked together constitutionally protected human rights and environmental protection through the use of Public Interest Litigation (PIL). Particular emphasis in this chapter is placed on the jurisprudence of the recently created National Green Tribunal (NGT) India, which is the new forum for environmental disputes.

The recognition that human survival depends in part upon the right to a healthy environment is gaining broader acceptance at the national, regional and international levels. Globally, there is increasing constitutional recognition and enforcement by the courts of the right to a healthy environment.[2] Additionally, the recognition by soft law instruments and treaties that are procedural, derivative or stand alone is increasingly based on an emerging understanding that environmental harm affects the right to life.[3]

In contrast, the right to development articulates the entitlement of the developing world.[4] Article 1 of the Declaration on the Right to Development states that 'the right to development is an inalienable human right by virtue of which every human person and all peoples are entitled to participate in, contribute to, and enjoy economic, social, cultural and political development, in which all human rights and fundamental freedoms can be fully realized'.[5] The right to development is a process encompassing a broad societal canvas of development that includes freedom from poverty, social deprivation and tyranny and determines the possibility of change and transformation and in particular ensures equity or the reduction of disparities.

---

[2] DR Boyd, *The Environmental Rights Revolution: A Global Study of Constitutions, Human Rights, and the Environment* (Vancouver, UBC Press 2012) 90.

[3] L Rajamani, 'The Increasing Currency and Relevance of Right-based Perspective in the International Negotiation on Climate Change' (2010) 22/3 *Journal of Environmental Law* 391, 407.

[4] A Lindroos, *The Right to Development* (University of Helsinki 1999) 29.

[5] Declaration on the Right to Development [4 December 1986], UNGA Res.41/128, Annex 41, Supp. No.53 at 186, UN Doc.A/41/53(DRD).

However, the convergence of the right to environment and the right to development create a paradox in India. India's population is gargantuan[6] as are its internal problems, including poverty,[7] inflation[8] and corruption.[9] To address these challenges, India embarked on a new path of economic growth in 1992. However, the absence of a strong state, and the incompetence or corruption of politicians and bureaucrats has reserved 'Shining India' for the select few: the rich and the powerful.[10] Skewed income and wealth distribution continues to increase the gap between rich and poor as a result of a non-inclusive growth pattern that undervalues low income levels and non-income factors that impact on the individual's quality of life.

India's economic growth agenda, however, often conflicts with human rights and protection of the environment. For example, the right to a healthy environment is endangered by industrialisation, modernisation and the impact of consumerism, resulting in air and water pollution, waste management problems and general environmental degradation. The latest World Health Organization (WHO) Report states that New Delhi, the capital of India, has the dirtiest air in India with an annual pollution average of 153 micrograms of small particulates, known as PM 2.5, per cubic meter, thereby placing city-dwellers at a very high risk of cancer, stroke and heart disease.[11]

---

[6] The current population of India is 1.27 billion people and is estimated to rise to 1.53 billion by 2030 (available at http://www.indiaonlinepages.com/population/india-current-population.html).

[7] The Government of India Planning Commission Press Note on Poverty Estimates 2009–2010, 19 March 2012. The Planning Commission of India has produced a poverty estimate of 29.8 per cent or 360 million people as being poor, although this figure is considered by some as inaccurate, being too low. See AV Kala, 'How to Read India's Poverty Stats' (July 2013) *The Wall Street Journal*; T Lahiri, 'Is India Fudging its Poverty Numbers?' (March 2012) *The Wall Street Journal India*; P Sainath, 'How to Beat Poverty in India' (March 2012) *Counter Punch*; S Desai, 'The Great Indian Poverty Game' (March 2012) *Business Standard*.

[8] The current rate of inflation in India is 7.08 per cent. Available at http://www.inflation.eu/inflation-rates/india/inflation-india.aspx (accessed 3.4.2015).

[9] India ranks 94th among 177 nations in the global corruption index released by Transparency International, available at http://www.dnaindia.com/india/report-global-corruption-perception-index-india-ranks-94th-among-177-nations-1929286.

[10] See United Nations Development Programme Human Development Report 2011, 'Sustainability and Equity: A Better Future for All 2011'.

[11] http://www.theguardian.com/world/2014/may/08/india-admits-delhi-matches-beijing-air-polllution-world-health-organisation-cities (accessed 24.3.2014).

The right to development is commonly perceived as being a reflection of powerful commercial vested interests rather than supporting equitable social welfare. Economic globalisation has created opportunities for investments, which in turn have resulted in the displacement of local people and damage to the local ecosystem.[12] Investment by foreign firms and industry, vigorously supported by Government, into sectors such as energy and mining has on occasions resulted in unsustainable development. Regulatory environmental procedures have been ignored in the race for economic returns. The result has left the poor and marginalised to suffer and die. Mining extraction in the Niyamgiri Hills, popularly known as 'Vedanta Mining', illustrates this point.[13]

Can these two human rights – to a healthy environment and to development – operate harmoniously within the new Indian economic strategy? This is a pressing question and has assumed heightened significance under the leadership of India's current Prime Minister, Narendra Modi, who is known for his ability to drive change and his promise to promote accelerated growth in Asia's third largest economy. Modi, known as a pro-business, market-oriented reformer, aims to boost India's growth to 8 to 10 per cent per annum in the next five years.[14] Resurgence in manufacturing such as car and car parts, pharmaceuticals, textiles and other goods; fast-tracking major infrastructure projects in power, roads, port and rail; improving the business environment; and encouraging foreign direct investment are some of the key projects that the Prime Minister aims to implement to bolster economic growth.

Resuscitating India's economy is necessary but it must not be detrimental to the environment. Striking a balance between human needs,

---

[12] See generally, L Zarsky, *Human Rights and the Environment: Conflicts and Norms in a Globalized World* (London, Earthscan Publishing 2002); D Aguirre, *The Human Right to Development in a Globalized World* (Aldershot, Ashgate 2008).

[13] http://www.survivalinternational.org/behindthelies/vedanta (accessed 19.3. 2014). Vedanta Resources is a London-based company. Its slogan is 'Creating Happiness', exercised through its social responsibility programmes. It was awarded the Golden Peacock Award for environmental management. Nevertheless, in India it has been repeatedly found to violate laws on environment protection and human rights. See 'Vedanta Resources lawsuit (re Dongria Kondh in Orissa)' (2014) *Business and Human Rights Resource Centre*; and the extensive resources linked to the report; http://business-humanrights.org/en/vedanta-resources-lawsuit-re-dongria-kondh-in-orissa (accessed 3.4.2015).

[14] N Mandhana, 'India's Modi Says his Government Will Push for Economic Development' (2014) *The Wall Street Journal* June; http://www.ozy.com/c-notes/what-modi-means-for-indian-biz/31684.article (accessed 15.6.2014).

economic growth and the environment is required to create an 'inclusive development' that improves people's well-being and capabilities for a fully functioning life.[15] In this context, the role of the Indian judiciary assumes seminal importance. The public has lost faith in politics as a consequence of corrupt and ineffective political leadership and bureaucratic failure and has increasingly relied on the courts to deliver social and economic justice. Some judicial decisions have resulted in groundbreaking new insights, whereas at other times there have been setbacks.

This chapter traces the successes and failures of PIL as a legal procedure to both promote economic activity and simultaneously to maintain the environment. The chapter thereafter focuses on the changes and benefits, both actual and expected, of the NGT. Thus, the chapter traces the jurisprudential environmental path from PIL in the Supreme Court along its journey to the NGT. Accordingly, after this Introduction the chapter is divided into three parts. Part 2 offers an account of the Indian courts' focus on PIL and the genesis of NGT. Part 3 presents and analyses the judicial attempts to redress environmental harms and protect human rights and associated legitimate interests by examining the two rights (the right to environment and the right to development): first, by engaging with the right to a healthy environment as a part of the right to life at the national level; and secondly, by examining the tensions concerning competing interests between the right to development on the one hand and safeguarding the rights of marginalised and disadvantaged poor and environmental protection on the other. The final part encapsulates the conclusion of this analysis.

---

[15] See generally, Amartya Sen and Martha Nussbaum who developed a 'capability approach', envisaging both the qualities and capabilities held by people and their ability to express and exercise those capabilities in a functioning life. A Sen, *Commodities and Capabilities* (Oxford, Oxford University Press 1999); M Nussbaum, *Frontiers of Justice: Disability, Nationality, Species Membership* (Cambridge MA, Harvard University Press 2006); M Nussbaum and A Sen, *The Quality of Life* (Oxford, Oxford University Press 1992).

## 2. JUDICIAL MECHANISMS: PUBLIC INTEREST LITIGATION AND THE NATIONAL GREEN TRIBUNAL

### 2.1 Public Interest Litigation

Environmental regulations in India aim to prevent environmental degradation and to protect and preserve natural resources. However, current regulations are inadequate as they fail to create substantive prohibitions on environmental destruction and also vest significant discretionary powers in administrative authorities. The slack performance of inadequately funded statutory bodies and enforcement agencies, political interference and a lack of will to tackle ensconced industrial and commercial interests contribute to poor environmental administration.[16] In *Indian Council for Enviro-Legal Action v Union of India*[17] the court declared that:

> if the mere enactment of laws relating to the protection of environment was to ensure a clean and pollution free environment, then, India would, perhaps, be the least polluted country in the world. But, this is not so. There are stated to be over 200 Central and State statutes which have at least some concern with environmental protection, either directly or indirectly. The plethora of such enactments has, unfortunately, not resulted in preventing environmental degradation which, on the contrary, has increased over the years.[18]

This general failure to prevent environmental degradation has resulted in the Indian judiciary earning public respect and credibility because of its innovative and creative efforts to protect ecology and the environment. The use of PIL as a broad-based, people-oriented approach that provides access to justice has become a 'wheel of transformation' for victims of environmental degradation.[19]

---

[16] South Asian Human Rights Documentation Centre, *Human Rights and Humanitarian Law: Developments in Indian and International Law* (Oxford, Oxford University Press 2008) 423; S Divan and A Rosencranz, *Environmental Law and Policy in India* (Oxford, Oxford University Press 2001) 2–3.

[17] 1996 (5) SCC 281.

[18] Ibid 293.

[19] GN Gill, 'Human Rights and the Environment in India: Access through Public Interest Litigation' (2012) 14 Environmental Law Review 201; SP Sathe, *Judicial Activism in India Transgressing Borders and Enforcing Limits* (Oxford, Oxford University Press 2002) 210.

PIL does not have the characteristics of traditional common law litigation, which involves the individualisation of disputes. PIL does not reflect the enforcement of the right of one individual against another, but rather, promotes and vindicates public interests that demand redress concerning the violation of the constitutional or legal rights of people, communities or society – particularly the poor, ignorant or those in a socially or economically disadvantaged position. PIL in India was created and implemented by judicial craftsmanship in order to forge new remedies and fashion new strategies by expounding and enforcing Indian constitutional mandates.[20]

Three constitutional mandates have played a significant role in underpinning environmental protection in India. The first is the fundamental right to life and liberty (Article 21). The second is the directive principle of state policy that protects and improves the environment and safeguards the nation's forests and its wildlife (Article 48 A). The final mandate is a fundamental duty upon citizens to protect and improve the natural environment, including forests, lakes, rivers and wildlife and to have compassion for living creatures (Article 51A (g)).

Concomitant with the rise of PIL, the Supreme Court of India has expanded the scope of constitutional provisions and combined human rights and the environment in order to build a new environmental jurisprudence. In *State of Uttaranchal v B.S. Chaufal*,[21] the court explained the need to entertain environmental PIL by stating that:

> the scale of injustice occurring on the Indian soil is catastrophic. Each day hundreds of thousands of factories are functioning without pollution control devices. Thousands of Indians go to mines and undertake hazardous work without proper safety protection. Every day millions of litres of untreated raw effluents are dumped into our rivers and millions of tons of hazardous waste are simply dumped on the earth. The environment has become so degraded that instead of nurturing us it is poisoning us. In this scenario, in a large number of cases, the Supreme Court intervened in the matter and issued innumerable directions.[22]

Environmental PIL is a product of the court's response to the inaction of the state and to the wrongful action of state agencies with regard to the

---

[20] See Divan and Rosencranz (n 16) at 133; MG Faure and AV Raja, 'Effectiveness of Environmental Public Interest Litigation in India: Determining the Key Variable' (2010) *Fordham Environmental Law Review* XXI 225; P Leelakrishanan, *Environmental Law in India* (Butterworth India 2005).

[21] (2010) 3 SCC 402.

[22] Ibid 437.

performance of their statutory duties which have resulted in endangering or in impairing the quality of life of people as guaranteed by Article 21 of the Constitution of India. *M.C. Mehta v Union of India*[23] (popularly known as the 'Ganga Pollution Tanneries Case') illustrates the point. A PIL was filed by M.C. Mehta (a well-known, public-spirited citizen) against the tanneries to restrain them from releasing trade effluents into the holy river Ganges until such time as they installed the necessary treatment plants in order to arrest water pollution. For several years, the tanneries were asked to take the necessary steps to establish primary treatment plants in order to prevent the flow of untreated waste water. The effluent discharged from a tannery is ten times more noxious than the domestic sewage water that flows into the river. The court ordered the closure of those tanneries that failed to take steps to establish primary treatment plants, observing:

> we feel that the tanneries at Jajmu, Kanpur cannot be allowed to continue to carry on the industrial activity … In cases of this nature this court may issue appropriate directions if it finds that the public nuisance or other wrongful act affecting or likely to affect the public is being committed and the statutory authorities who are charged with the duty to prevent it are not taking adequate steps to rectify the grievance. We are conscious that closure of tanneries may bring unemployment, loss of revenue but life, health and ecology have greater importance to the people.[24]

The expansion of environmental jurisprudence is a result of a collaborative effort built on innovative substantive and procedural approaches in order to make effective human rights and promote environmental discourse. The substantive approaches cover the extension of the right to life to include the right to a wholesome environment and the derivative application of international environmental law – precautionary and polluter pays principles, inter-generational equity and the public trust doctrine.[25] For example, in *Intellectual Forum, Tirupathi v State of Andhra Pradesh*[26] the Supreme Court recognised inter-generational equity and the public trust doctrine as an essential part of the right to life

---

[23] AIR 1988 SC 1037.

[24] Ibid 1047–48.

[25] L Rajamani, 'Public Interest Litigation in India: Exploring Issues of Access, Participation, Equity, Effectiveness and Sustainability' (2007) 19/3 *Journal of Environmental Law* 293; G Sahu, 'Implication of Indian Supreme Courts Innovation for Environmental Jurisprudence' (2008) 4/1 *Law, Environmental and Development Journal* 3–19.

[26] AIR 2006 SC 1350.

and observed that 'all human beings have a fundamental right to a healthy environment commensurate with their well-being ... ensuring that natural resources are conserved and preserved in such a way that the present as well as future generation are aware of them equally'.[27] Similarly, in *Vellore Citizen's Welfare Forum v Union of India*[28] the court observed that 'the precautionary and polluter pays principles have been accepted as a part of the law of the land. Article 21 of Constitution guarantees protection of life and liberty'.[29]

The relevant procedural innovations aim to improve access to justice to affected people as '... procedural prescriptions are handmaidens, not mistresses, of justice and failure of fair play is the spirit in which courts must view (procession) deviances'.[30] The modification of traditional *locus standi* has provided both the opportunity to, and the ability for, public spirited citizens and organizations to use the courts as a route to seek remedies through collective proceedings instead of being driven into an expensive plurality of litigations, thereby, affirming participative justice. Traditional *locus standi* has been modified in two ways: by representative and citizen standing. The court has emphasised that any member of the public having sufficient interest may be allowed to initiate the legal process in order to assert diffused and meta-individual rights in response to environmental problems.[31]

Representative standing allows any member of the public, acting *bona fide*, to advance claims against the violation of human rights and against denials of environmental justice to victims who by reason of poverty, helplessness, disability or socially or economically disadvantaged position could not approach the court for the judicial enforcement of their fundamental rights. The tension between tribal rights, environmental rights and the exigencies of development for the protection of forest lands and livelihood are cases covered under representative standing. Any citizen, acting *bona fide* and with sufficient interest for the purpose of redressing public injury, enforcing public duty, protecting social, collective, diffused rights and interests or vindicating the public interest can be accorded 'citizen standing'. Thus, the liberal interpretation of *locus standi* has removed traditional procedural shackles in order to enable the hearing of petitions for remedying hardships resulting from a gross

---

27 Ibid para 84.

28 (1997) 3 SCC 715.

29 Ibid 720.

30 *Mumbai Kamgar Sabha v Abdulbhai Faizullabhai* AIR 1976 SC 1455.

31 *RLEK v State of Uttar Pradesh* AIR 1985 SC 652; Judgment of 19 December 1996.

violation of fundamental rights and brought by a group or class action, or when basic human rights are invaded, or where complaints of such acts have shocked the judicial conscience.[32]

The power to appoint an expert committee or special commission comprised of environmental experts is another novel procedure used by the court to resolve environmental disputes. These appointments are made in order to construct a comprehensive account of relevant facts, receive expert testimony in cases involving complex or scientific issues and ensure the continuous supervision of prospective judicial orders.[33]

Notwithstanding the benefits of the above-mentioned innovative procedures, the judiciary is not without critics who decry the relaxation of *locus standi*, which has resulted in adding to the already staggering number of pending cases.[34] The environmental PIL was conceived to promote open access to potential litigants at affordable levels of cost and risk. However, now it can be both expensive and time-consuming. For example, in the Delhi Vehicular Pollution case, the original writ petition was filed in 1985. The case remains active to this day, although many interim directions and orders have been passed.[35]

However, a recent empirical study by Sahu[36] argues that the earlier general findings by Galanter in 1974 do not apply to later Indian environmental cases. Galanter established that the 'haves' are more successful than the 'have nots' when it comes to access to justice in India.[37] Sahu documents that the environmental judgments delivered between 1980 and 2010 reflect the fact that weaker parties, including NGOs and activists, have been relatively successful as court petitioners

---

[32] See Gill (n 19) at 205–206; U Baxi, The Avatars of Indian Judicial Activism: Explorations in the Geographies of (In) Justice' in SK Verma and K Kumar (eds), *Fifty Years of the Supreme Court of India: Its Grasp and Reach* (Oxford, Oxford University Press 2000); P Bhushan, 'Supreme Court and PIL' (2004) 39/18 *Economic & Political Weekly*, 1770–74.

[33] See Divan and Rosencranz (n 16) 143; see Sahu (n 25) 11; and also G Sahu, 'Implementation of Environmental Judgments in Context: A Comparative Analysis of Dahanu Thermal Power Plant Pollution Case in Maharashtra and Vellore Leather Industrial Pollution Case in Tamil Nadu' (2010) 6/3 *Law, Environment and Development Journal* 335–53.

[34] See Rajamani (n 25).

[35] See Gill (n 19).

[36] G Sahu, 'Why the Underdogs Came Out Ahead: An Analysis of the Supreme Court's Environmental Judgments, 1980–2010' xlix/4 (2014) *Economic and Political Weekly* 52–57.

[37] M Galanter, 'Why the "Haves" Come Out Ahead: Speculations on the Limits of Legal Change' (1974) 9/1 *Law and Society Review* 95–160.

when bringing actions against the Union of India, state governments and industry. The courts are not meant exclusively for the 'top dogs' (federal and state governments and the business groups) but also exist for the 'underdogs' (individuals, communities, NGOs). Social and economic justice is the essence of the Indian Constitution, which in turn places a duty on the judiciary to protect the rights of every citizen, and particularly of the vulnerable sections of society.

However, according to Sahu, on matters of public infrastructure such as the construction of dams and power stations, the court has on occasions moved from its strong environmental position in order to allow economic development on the basis of national economic needs.[38] Several judges known for their pro-environment positions have adopted a position of non-confrontation with the state in regard to such petitions. They have allowed such projects to proceed despite social and environmental concerns and negative consequences. This is problematic, because large infrastructure projects involve two highly relevant and important trends.[39] First, environmental protection clashes with powerful vested commercial and corporate interests. Secondly, the distributional impact on the affected poor and marginalised is both significant and disproportionate.

Sahu's data analysis is illustrated by two contrasting judicial decisions concerning environmental bias. In 1987, in *Banwasi Seva Ashram v State of Uttar Pradesh*,[40] the Supreme Court accepted a letter written to the court by Banwasi Seva Ashram, a non-governmental organisation working towards protecting the rights of tribal forest dwellers ('*adivasis*' and other 'backward' people).[41] The main grievance of the NGO was acquisition proceedings initiated by the state government in order to locate a super thermal power plant of the National Thermal Power Corporation (NTPC) in the reserved forest area and the displacement of the *adivasis* from the forest area that had been their habitat and means of

---

38 Ibid 55.

39 PR Bhushan, 'Misplaced Priorities and Class Bias of the Judiciary' (2009) 44/14 *Economic & Political Weekly* 32–37; V Gauri, 'India in Transition, Public Interest Litigation in India: Time for an Audit', available at http://casi.ssc.upenn.edu/iit/gauri (accessed 14.1.2014).

40 AIR 1987 SC 374.

41 'Backward Class' is a collective term used by the Government of India to classify castes that are educationally and socially disadvantaged. It is one of several official classifications of the population of India other than the Scheduled Castes and Scheduled Tribes in the lists prepared by the Government of India; http://socialjustice.nic.in/aboutdivision4.php.

livelihood for generations. It was alleged by the NGO that the forest officers (the authorities responsible for protecting the rights of the tribal people) were involved in encroaching upon land and obstructing the free movement of the tribal people. The Supreme Court acknowledged the PIL by observing that:

> forests are a much wanted national asset. On account of the depletion thereof ecology has been disturbed; climate has undergone a major change and rains have become scanty. These have long-term adverse effects on the national economy and also on the living process. Further, it is common knowledge that the *adivasis* and other backward people living within the jungle used the forest area as their habitat ... for generations they had been using the jungles for collecting the requirement for their livelihood – fruits, vegetables, fodder, flowers, timbers, animals and fuel wood.[42]

At the same time the court recognised that for industrial growth and the provision of improved living facilities there is a great demand for energy, particularly electricity. Nevertheless, the Supreme Court, in protectionist mode, passed stringent and detailed directions safeguarding and protecting the forests and the interests of *adivasis* by permitting the acquisition of land only after NTPC agreed to provide facilities, including rehabilitation measures, monetary compensation for crops and land and legal aid to the ousted forest dwellers.

The more recent Commonwealth Games case illustrates the judicial trend of favouring 'infrastructure development' by condoning misplaced priorities and erroneous decisions of local authorities.[43] The court was faced with choosing between a controversially sited urban development or providing ecological protection.[44] The construction of the Commonwealth Games Village (CWGV) in Delhi on a legally protected floodplain of the river Yamuna was challenged before the Supreme Court on the ground that it was in violation of the Environmental Protection Agency regulations which stipulated that no permanent structures could be built on the river banks. The hosting of the Commonwealth Games for India in 2010 was a matter of national pride providing opportunities for economic profit, urban regeneration and international exposure and also to showcase Delhi as a 'world class city' and India as Asia's new superpower.

---

42 Ibid 376.

43 GN Gill, 'Environmental Protection and Development Interests: A Case Study of the River Yamuna and the Commonwealth Games, Delhi 2010' (2014) 6/1–2 *The International Journal of Law in the Built Environment [Special Issue on Environmental Law]* 69–90.

44 *DDA v Rajendra Singh* AIR 2010 SC 2516.

The Commonwealth Games development reality was disturbingly different, however. The social and environmental impact was considerable. The displacement of the 'invisible common poor' from their homes to create building space was large scale: over 100,000 families were displaced due to Commonwealth Games related projects. Beggars and the homeless were arrested and arbitrarily detained in the 'no tolerance zone'.[45] The Delhi Minister for Social Welfare, Mangat Ram Singhal declared that '[b]eggars are a nuisance and begging has to be stopped. When we make Delhi a world class city, it will be compared with other class capitals. One does not come across beggars in other countries. Why should there be beggars in Delhi? We must make the city free of them.'[46] The environmental impacts of constructing the CWGV on the floodplain of the Yamuna river, the 'lifeline of Delhi and a source of major supply to the city'[47] impaired the water recharging function of the floodplain ecosystem and caused damage to natural resources, migratory avifauna and to the wetland to which the Yamuna floodplain area belonged. Despite these negative impacts, the Supreme Court did not stop the construction and declared that the site was not located on the floodplain, despite the concerned observation of the Delhi High Court that the matter should be further investigated by means of an expert committee. As soon as the Games concluded, the then Prime Minister of India, Manmohan Singh, announced a High Level Committee under V.K. Shunglu to examine issues concerning the Commonwealth Games with special reference to the CWGV. As a result of the investigation, various Games officials were charged with corrupt practices and the selection of the site was severely criticised.[48] The former Environment and Forest Minister, Jairam Ramesh was asked in 2011 whether environmental clearance should have been granted for the CWGV. Ramesh stated that the clearance should not have been given as he felt that the Yamuna river bed was devastated by its construction. Was the Village located on the

---

[45] Housing and Land Rights Network Report [2010], 'The 2010 Commonwealth Games: Whose Wealth, Whose Commons?', Housing and Land Rights Network – South Asia Regional Programme Habitat International Coalition, New Delhi India 3.

[46] AA Mahaprashasta, 'War on Beggars' (2010) *Frontline, The Hindu* 19 June–2 July 27/13.

[47] M Misra, 'Dreaming of a Blue Yamuna', in B Chaturvedi, *Finding Delhi: Loss and Renewal in the Mega City* (New Delhi, Penguin Viking 2010) 71.

[48] The High Level Committee Report on Commonwealth Games [2011] Government of India.

floodplain?[49] The Yamuna provided its own answer to the question in August and September 2010 when the Village was flooded by monsoon rain. This case illustrates the tensions that the court faces when subjected to powerful political and economic interests, both in the private and public sectors. The court's minimalist responses at key points in the planning and construction of the Village were ultimately to the detriment of the river Yamuna and also to India's environmental jurisprudence.

Additionally, PIL has been affected by complex scientific and technical issues relating to the environment. Appointing or referring matters to expert committees or commissions not only exposes the court's dependency on such bodies but may also result in contradictory or divided findings that place further evidential choices on the judiciary. The composition and functioning of such court appointed committees and commissions often bear witness to a strong inclination to forward particular departmental agendas (for example conservation) without taking a holistic view of an environmental problem. The creation of such bodies may also result in the creation of a parallel power structure within the governance framework.[50]

Thus, although the Supreme Court created a procedure that allowed indigents and concerned citizens to access the courts via PIL, the process did not prove to be a 'magic bullet'. As with all processes it had its weaknesses and limitations. Innovation and change were needed. This occurred through the establishment of the National Green Tribunal.

## 2.2 The National Green Tribunal: A New Environmental Forum

The judicial genesis of environmental courts is found in three major judgments of the Supreme Court of India. In *M.C. Mehta v Union of India*[51] the Supreme Court stated that as environmental cases frequently involve assessment of scientific data, it was desirable to set up environmental courts on a regional basis with a legally qualified judge and two experts to undertake relevant adjudication. Another influential judgment was the *Indian Council for Enviro-Legal Action v Union of India*,[52] where the Supreme Court observed that:

---

49 *The Hindu* (2011), 'River Regulation Zone Coming: Jairam' January 7, available at http://www.thehindu.com/news/national/river-regulation-zone-coming-jairam/article1063315.ece (accessed 12.11.2013).

50 See Sahu (n 25) 13; BN Srikrishna, 'Judicial Activism – Judges as Social Engineers, Skinning a Cat' (2005) 8 SCC J 3.

51 AIR 1987 SC 965.

52 (1996) 3 SCC 212.

> the suggestion for the establishment of environment courts is a commendable one. The experience shows that the prosecutions launched in ordinary criminal courts under the provisions of Water Act, Air Act and Environment Act never reach their conclusion either because of the workload in those courts or because there is no proper appreciation of the significance of the environment matters on the part of those in charge of conducting those cases. Moreover, any orders passed by the authorities under Water, Air or Environment Acts are immediately questioned by the industries in courts. Those proceedings take years and years to reach conclusion. Very often, interim orders are granted meanwhile which effectively disable the authorities from ensuring the implementation of their orders. All these point to the need for creating environment courts which alone should be empowered to deal with all matters, civil and criminal, relating to the environment. These courts should be manned by legally trained persons/judicial officers and should be allowed to adopt summary procedures.[53]

Again, in the judgment of *A. P. Pollution Control Board v Professor M V Nayudu*,[54] the Supreme Court referred to the need for environmental courts. Such courts would have the benefit of expert advice from technically qualified environmental scientists as part of the judicial process. It was suggested that the Law Commission of India should examine this possibility.

These three judgments were a result of concern within the Supreme Court about complexity and uncertainty underpinning the scientific evidence presented to the court. Such evidence generated tensions between the fears expressed by claimants and the assurances given by the defendants.

Following the observations made by the Supreme Court of India, the Law Commission of India in its One Hundred and Eighty Sixth Report,[55] 'Proposal to Constitute Environment Courts', strongly advocated the establishment of 'Environment Courts' staffed exclusively by persons with judicial or legal experience but supported by persons having scientific qualifications and experience in the environmental field. The Report sought accessible, open and speedy justice. In addition, it was suggested that the establishment of environmental courts would reduce the burden on the High Courts and on the Supreme Court, given the growth of public interest litigation in India involving complex and technical environmental issues.

---

53 Ibid 252.

54 (1999) 2 SCC 718 and (2001) 2 SCC 62.

55 The Law Commission of India (2003) One Hundred and Eighty Sixth Report, 8–9.

Accordingly, the Indian Parliament passed the National Green Tribunal Act in June 2010.[56] The Act provides for the establishment of a National Green Tribunal (NGT). The Tribunal decides cases relating to environmental protection and to the conservation of forests and other natural resources, including enforcement of any legal right relating to the environment and gives relief and compensation for damages to persons and property. The NGT was established on 18 October 2010 and commenced hearings on 5 May 2011,[57] with New Delhi selected as the site for the principal bench, although Bhopal was mooted earlier in acknowledgement of the environmental industrial disaster of 1984. The then Environment Minister, made a statement in the Indian Parliament in 2010 that:

> the main bench of the tribunal will be in Bhopal. This way the government and parliament could show some sensitivity to the people of Bhopal, the site of the worst industrial disaster. We can never obliterate that tragedy from our memories but by setting the national green tribunal in Bhopal, I think we would send a signal that we mean business. A circuit approach would be followed to enable access for people. The court will go to the people. People would not come to the court.[58]

The principal bench in New Delhi covers northern India; while the regional benches find their home in Bhopal for the central zone; Chennai for south India; Pune for the western territory and Kolkata is responsible for the eastern region.[59]

A remarkable feature of the NGT is its composition. The NGT is a specialised body where the decision makers hold relevant qualifications and appropriate experience, some in law and others as scientific experts with environmental knowledge.[60] The judicial and scientific bench members act as joint decision makers of equal standing, thereby promoting multi-disciplinary decision making. The expertise of the NGT is reflected in the enhanced quality of the Tribunal's decisions. The NGT benches not

---

[56] The National Green Tribunal Act, 2010, The Gazette of India Extraordinary (No. 19 of 2010); see GN Gill, 'A Green Tribunal for India' (2010) 22/3 *Journal of Environmental Law* 461–74.

[57] Ministry of Environment and Forests Notification 5th May 2011 S.O.1003 E.

[58] Statement made by Jairam Ramesh, Former Minister of State of the Ministry of Environment and Forests in the Indian Parliament on 5th May 2010, The Rajya Sabha Debates 246.

[59] Ministry of Environment and Forests Notification 17th August 2011 S.O.1908 E.

[60] Section 5 NGT Act 2010.

only decide matters based on scientific evidence but also make positive policy suggestions to improve environmental management.[61]

The NGT has wide jurisdiction in relation to environmental matters. The Tribunal has both original and appellate jurisdiction under the National Green Tribunal Act 2010. The NGT's original jurisdiction is exercised in civil cases in relation to a substantial question relating to the environment.[62] This includes enforcement of any legal right relating to the environment and addressing such questions as arise out of the implementation of the enactments specified in Schedule 1 of the Act.[63] The wording 'all civil cases' includes all legal proceedings except criminal cases, which are governed by the provisions of the Criminal Procedure Code.[64] Section 2(m) of the National Green Tribunal Act classifies a 'substantial question on environment' under two heads: firstly, where there is a direct violation of a statutory duty or environmental obligation which is likely to affect the community or where the gravity of damage to the environment or property is substantial or the damage to public health is broadly measurable; secondly, where the environmental consequences relate to a specific activity or a point source of pollution.

Under section 16 of the National Green Tribunal Act 2010, the Tribunal has appellate jurisdiction regarding the orders or decisions under the enactments specified in Schedule I.[65] Any aggrieved person has the right to appeal against such a decision or direction. It is important to note that the definition of an aggrieved person in environmental matters has been given a liberal construction, thereby giving a voice to human rights activists. In *Vimal Bhai v Ministry of Environment and Forests*,[66] the NGT bench stated that 'any person whether he is a resident of that particular area or not whether he is aggrieved and/or injured or not, can

---

[61] In *Vigyan Arogya Sanstha v Ministry of Environment and Forests*, Judgment dated September 20 2011, the NGT prescribed national standards as to permissible levels of nuclear radiation in residential, industrial and ecologically sensitive areas of India and synchronised the commissioning of the thermal power project with that of a sewage waste water treatment plant.

[62] Section 14(1) NGT Act 2010.

[63] The enactments in Schedule 1 include: The Water Act 1974; The Water Cess Act 1977; The Forests (Conservation) Act 1980; The Air Act 1981; The Environment Protection Act 1986; The Public Liability Insurance Act 1981 and The Biological Diversity Act 2002.

[64] *MP Pollution Control Board v Commissioner Municipal Corporation Bhopal*, Judgment dated August 8 2013.

[65] See n 61.

[66] Judgment dated December 14 2011. However, the Supreme Court of India in its March 10, 2014 order stayed the 2012 order as it requires reconsideration.

approach this tribunal. In such situations, it is of course necessary to review the credentials of applicants/appellants as to their true intentions and motives'. The bench, in its liberal interpretation, was guided by two reasons: first, the frequent inability of persons living in the area or vicinity of the proposed project to understand the intrinsic scientific details and the effects of the ultimate project and any disaster it may cause: hence the right for any citizen to approach the tribunal regardless of whether or not he/she is directly affected by a developmental project or whether he/she is a resident of the affected area or not; secondly, the subservience of statutory provisions of the National Green Tribunal Act 2010 to the constitutional mandate of Article 51A (g) giving every citizen a fundamental duty to protect and improve the natural environment. More recently, in *Betty C Alvares v State of Goa*,[67] the NGT has gone a step further and opened its doors globally to each and every resident person, including incorporated bodies, that consider themselves 'aggrieved' within India.

The establishment of the NGT provoked the Supreme Court of India to review its PIL environmental caseload and its limited environmental expertise. Consequently, in 2012, the Supreme Court in a PIL case transferred all environmental cases, both active and prospective, to the NGT.[68] Further, the Supreme Court advised the High Courts, at their discretion, to transfer to the Tribunal those environmental cases filed and pending prior to the coming into force of the NGT Act.

## 3. SHAPING AND STRENGTHENING THE HUMAN RIGHTS AND ENVIRONMENTAL PROTECTION DISCOURSE: AN OVERVIEW OF THE JUDICIAL APPROACH

### 3.1 The Right to a Healthy Environment: Towards Dignity and Life

Henry Shue[69] has pioneered the use of the 'linkage argument' to defend the right to subsistence: 'no one can fully … enjoy any right that is

---

[67] Judgment dated February 14 2014.

[68] *Bhopal Gas Peedith Mahila Udyog Sangathan v Union of India,* Order dated 9 August 2012. However, in *Adarsh Cooperative Housing Society Ltd v Union of India*, Order dated 10 March 2014, the Supreme Court of India has stayed its own order by which it had transferred all environmental cases from high courts in the country to the National Green Tribunal.

[69] H Shue, *Basic Rights* (Princeton NJ, Princeton University Press 1996) 24–25.

supposedly protected by society if he or she lacks the essentials for a reasonably healthy and active life'. Shue has defined subsistence as meaning 'unpolluted air, unpolluted water, adequate food, clothing, adequate shelter and minimal preventive care'.[70] Orend's definition is similar: 'material subsistence means having secure access to those resources one requires to meet one's biological needs – notably a minimal level of nutritious food, clean water, fresh air, some clothing and shelter, and basic preventative health care'.[71]

The inter-linkage between the right to life and a healthy environment is concerned with ensuring the conditions, negative and/or positive, of a minimally good life. The normative focus is 'on the individual – in the dignity and worth of the human person'.[72] The Supreme Court of India has developed its case law for human rights and environmental protection by providing an expansive interpretation to the term 'life' under Article 21 of the Constitution of India. It includes not only physical existence but also the quality of life. In *Francis Coralie v Delhi*,[73] Justice Bhagwati stated: 'We think that the right to life includes the right to live with human dignity and all that goes along with it, namely, the bare necessities of life such as adequate nutrition, clothing and shelter over the head and facilities for reading, writing, and expressing oneself in diverse forms'.[74] Subsequently, the Court recognised in *Virender Gaur* v *State of Haryana*[75] that a healthy environment is one free from environmental pollution. The Court observed:

> Article 21 protects the right to life as a fundamental right. Enjoyment of life … including the right to live with human dignity encompasses within its ambit, the protection and preservation of the environment, ecological balance free from pollution of air and water, sanitation, without which life cannot be enjoyed. Any contra acts or actions would cause environmental pollution. Environmental, ecological, air and water pollution, etc., should be regarded as amounting to a violation of Article 21. Therefore, a hygienic environment is an integral facet of the right to a healthy life and it would be impossible to live with human dignity without a human and healthy environment. … There is a constitutional imperative on the State Government and the municipalities, not only to ensure and safeguard a proper environment but also an imperative

---

70 Ibid.

71 B Orend, *Human Rights: Concept and Context* (Peterborough ON, Broadview Press 2001); JW Nickel, *Making Sense of Human Rights* (Oxford, Blackwell Publishing 2007) 139.

72 See Rajamani (n 25) 414.

73 AIR 1981 SC 746.

74 Ibid 753.

75 (1995) 2 SCC 577.

> duty to take adequate measures to promote, protect and improve both the man made and the natural environment.[76]

The preamble of the NGT Act 2010 also recognises the right to a healthy environment as part of the right to life. Principles underpinning international environmental law, namely sustainable development, precautionary and polluter pays principles, have been absorbed into the right to a healthy environment and applied by the NGT.[77]

A healthy environment is thus emphatically confirmed as part of the right to life in India, and was applied by the NGT in the case of *Supreme Court Group Housing Society Through its Secretary v All India Panchayat Parishad*.[78] The main issue was the ineffective implementation of the provision of the Noise Pollution (Regulation and Control) Rules 2000. The use of loudspeakers, music and public address systems during weddings, receptions and parties until late at night, without obtaining permission from concerned authorities, disturbed and affected the residents living in the area. It was argued that noise pollution affected the health and disturbed the sleep of the residents, particularly infants and the aged, and not only caused annoyance but also led to significant adverse health disorders including higher blood pressure, hearing impairment and neurological disorders. Children are most susceptible to noise pollution which may impair the development of their mental capacity.

The NGT decided that the Noise Pollution (Regulation and Control) Rules 2000 were followed more in breach than compliance. The Tribunal observed that:

> the effect of noise on health is a matter which has not yet received the full attention which it deserves. Pollution, being the wrongful contamination of the environment which causes material injury to the right of an individual[,] [noise] can … be regarded as a pollutant because it contaminates environment with high decibel noise intensity, causes nuisance and affects the health of a person and would therefore offend Article 21 of the Constitution if it exceeds a reasonable limit.[79]

Keeping public health in mind, the Tribunal directed the authorities, especially the sub-divisional magistrate and local police, to ensure that the implementation of the Noise Pollution Rules 2000 is strictly observed, failure in respect of which should be viewed as a serious

---

[76] Ibid 580–81.
[77] Section 20 NGT Act 2010.
[78] Judgment dated December 18 2012.
[79] Ibid para 6.

violation of law justifying strict action to be taken against the officer, including disciplinary action. The Tribunal also required a proper strategy and suggested an action plan involving various departments to mitigate noise pollution problems in the city. These included: establishing and running a 24/7 call centre to lodge citizen complaints; the requirement for complaints to be attended by the police and a sub-divisional magistrate within 24 hours; providing adequate noise meters to all police stations to enable them to check the noise levels emanating from the various sources; periodical surveys in different parts of Delhi, especially in sensitive areas such as hospitals, educational institutions, courts and others, and those located close to heavy traffic intersections; and conducting educational and awareness noise pollution programmes.

It is submitted that the judgment passed by the NGT is well intentioned and policy orientated – but that realities on the ground reveal considerable shortcomings and continuing challenges. Public health is defined in a tangible form to include a safe, healthy and wholesome environmental milieu, free from pollution. However, the variability of implementation of noise pollution orders is challenging in India as the 'quantification of safe and healthy environment is difficult',[80] if not impossible. Establishing and adopting quality standards demands extensive research and debate involving public participation based upon impact studies. Resorting to approximation based upon pollution levels, availability of pollution control technology, technological capabilities, the efficacy of existing legal norms, and expense, are prime considerations for setting quality standards – an exercise based upon relative experience and not absolute targets.[81] Scientific evidence reveals that noise pollution is gripping Delhi and triggering Presbycusis (age-related hearing loss) at the age of 60, which normally sets in after the age of 75 years.[82] The

---

[80] A Kiss and D Shelton, *International Environmental Law* (United Nations Environment Programme 2003) 393.

[81] A Agarwal, *Slow Murder: The Deadly Story of Vehicular Pollution in India* (Centre for Science and Environment 1996) 51.

[82] N Chandra, 'New Study Warns "Dangerous" Levels of Noise Pollution in New Delhi are Causing Age-related Hearing Loss 15 Years Earlier than Normal' (2013) *Mail Online India* June, available at http://www.dailymail.co.uk/indiahome/indianews/article-2334486/Delhi-uncivilised-city-Dangerous-levels-noise-pollution-Capital-causing-onset-age-related-hearing-loss-early-60-new-study-warns.html (accessed 19.6.2014); N Lalchandani, 'Rules on Paper, Delhi Stays Noisy' (2013) *The Times of India* January, available at http://timesofindia.indiatimes.com/city/delhi/Rules-on-paper-Delhi-stays-noisy/articleshow/18190784.cms (accessed 12.4.2014); J Nandi, 'Most Parts of Delhi in Grip of Noise Pollution'

rules and orders remain on paper as the policies do not focus on regulating noise and noise-monitoring stations are inefficient.[83] The quintessential Delhi-ites' spirit is loud and unaware of the health hazards associated with noise.

Within the urban environment, extremely unhygienic and unsanitary conditions resulting from non-performance of statutory duties by the municipality also offend the right to life: the maintenance of health and the preservation of sanitation and the environment falls within Article 21, as failure in these duties adversely affects the life of the citizen because of the hazards created. The case of *Rayons-Enlightening Humanity v Union of India*[84] illustrates this issue. An original application was filed by Rayons-Enlightening Humanity, a non-governmental organisation against the operation of a municipal solid waste management plant (MSWM plant) at Rzazu-Paraspur, Bareilly. The applicant opposed the operation of the project as being illegal, not in accordance with environmental clearance regulations and based on arbitrary decisions of the municipal authorities. The plant was established in an area very close to residences, educational institutions, hospitals and water bodies, thereby resulting in environmental harm. The major part of the plant was in the open air and its basin pits had not been prepared in compliance with the Schedule to the Municipal Solid Waste Rules. The NGT allowed the application and observed that:

> the site in which the plant is located, is bound to cause pollution of ground water, which is relatively at a higher level, by leaches. This is inevitable especially in the rainy season. The municipal solid waste was been dumped in the open area at the site without any impermeable membrane lining. Therefore, the contaminated water is bound to seep into the underground water and even affect the adjoining water bodies apart from affecting irrigation water. The foul smell arising from the dumping at the site is bound to pollute the air quality of the area. This bad smell is bound to affect the health of the residents in the vicinity of the site and because of pollution they would be exposed to diseases like asthma, emphysema or even cancer. Thus, the adverse effects of permitting the plant to carry on its activities at the site in question are bound to cause irretrievable damage to public health and environment.[85]

---

(2013) *The Times of India* November, available at http://timesofindia.indiatimes.com/home/environment/pollution/Most-parts-of-Delhi-in-grip-of-noise-pollution/articleshow/25717955.cms (accessed 10.2.2014).

[83] Ibid.

[84] Judgment dated July 18 2013.

[85] Ibid para 47.

Accordingly, the Tribunal applied the precautionary principle and ordered the immediate closure of the municipal solid waste management plant. The environmental ill-effects and considerations of public health outweighed permitting the MSWM plant to continue operations at the village site.

Similarly, in *Paryavaran Mitra v Gujarat Pollution Control Board*,[86] the NGT applied the polluter pays principle and ordered a private firm M/s Hanjer Biotech, which managed municipal solid waste, to pay compensation of Rupees 25 Lakhs (£25,000) for its failure to handle the waste in a proper manner at the Nakravadi village in Rajkot municipality, Gujarat. The villagers faced serious health issues, contamination of underground water and an adverse impact on their agricultural land as the waste was not properly stored or recycled. The problem became more severe during the monsoon, resulting in untreated water flowing into the village.

Although the NGT is focused on effective environmental management, it is clear that apathetic and corrupt enforcement procedures may blunt its rulings. For example, managing waste is a major concern for India and its cities. Delhi produces 8,000 metric tons of waste every day,[87] followed by Mumbai at 6,500[88] and Ahmedabad[89] and Bengaluru[90] at 4,000. Massive mounds of stinking municipal solid waste can be found in the back streets of every city. The Union of India through the Ministry of Environment and Forests gazetted the Municipal Solid Waste (Management and Handling) Rules 2000, making the local administration of cities responsible for collecting, segregating and scientifically disposing of waste. In reality, the rules are frequently violated. It has been alleged that there exists a well-oiled nexus of 'mafia type business' between elected representatives and private contractors responsible for collecting waste. Faking bills, falsifying the number of trips and the amount of waste deposited at the dumpsite contribute to irregularities and illegalities.[91] Private contractors are usually the relatives of the elected

---

[86] Judgment dated December 20 2013.

[87] http://www.chintan-india.org/documents/fact_sheets/chintan_waste_fact_sheet.pdf (accessed 3.4.2015).

[88] www.mcgm.gov.in/irj/portal/anonymous/qlcleanover (accessed 11.5.2014).

[89] http://timesofindia.indiatimes.com/city/ahmedabad/Garbage-mounds-Spreading-urbaneyesores/articleshow/36074988.cms (accessed 16.6.2014).

[90] http://www.downtoearth.org.in/content/lessons-two-cities (accessed 14.5.2014).

[91] http://www.livemint.com/Politics/6c9QI6oD8fZ6ydfXhdVDCM/Managing-waste-irregularities-a-tough-call-for-India8217.html (accessed 16.5.2015);

representatives and are unaccountable for their acts. For example, the Ahmedabad Municipal Corporation unearthed a garbage scam at the Pirana landfill site where private contractors were dumping mud and stones in place of garbage and claiming payments for the entire weight deposited in the landfill.[92]

A review of the above-mentioned judgments reflects the judicial approach of preserving the jurisprudential link between life and a healthy environment. The NGT has interpreted serious and complex harms to the environment or to human health as being not individual-centric but socio-centric, thereby acknowledging the larger public interest. Thus, the right of an individual to a healthy and clean environment including air, water, soil and a noise-free environment is of paramount consideration. Accordingly, it is recommended that resources should be utilised in a planned manner to support a healthy existence.

## 3.2 The Right to Development: Incorporating Sustainability

Striking a balance between the environment and development continues to be both complex and contentious. In India and other countries with emerging economies, the word 'development' is interpreted as 'economic betterment' but often this is a misnomer. This dichotomy is to be found when environmental issues and the rights of the tribal and backward classes conflict with state development plans and multinational operations, which in many cases are supported by powerful local elites capable of manipulating political, legal and judicial systems. The state often advances its economic agenda without adequate participation by ordinary people in the development process. India is a democratic state but its decisions have on occasions ignored the well-being of the people.

Today, there is a growing recognition among Indian policy makers and implementers of the need to recognise the rights of tribal and indigenous people to land and resources and the importance of traditional knowledge for the ecologically sustainable use, management and preservation of ecosystems.[93] Nonetheless, conflicts between tribal poor and powerful

---

*dnaSyndication*, 'HC Pulls Up Government over Mafia Nexus' June 2014, available at http://dnasyndication.com/showarticlerss.aspx?nid=641PWB3tuoqPagLpt9dgw1nfZ5P2eOVVKBLkwN7tFgo= (accessed 21.6.2014).

[92] See above n 82.

[93] The Scheduled Tribes and Other Traditional Forest Dwellers (Recognition of Forest Rights) Act, 2006; The Scheduled Tribes and Other Traditional Forest Dwellers (Recognition of Forest Rights) Rules, 2007 and 2012; The Panchayats Extension to Schedule Areas 6 (PESA) 1996.

rich occur regularly. The role of NGT in this context is laudable as its judgments recognise that 'development' is the essence of any pragmatic and progressive society based upon a 'balancing act' of not only supporting economic benefits but also encompassing the guarantee of protecting the environment and the fundamental human rights of the poor. The 'balancing act' includes the full spectrum of civil, political, cultural, economic and social process in a holistic manner to conserve ecology and to improve the tribal and marginalised people's well-being. The following judgments illustrate the nexus between development, ecological concerns and the welfare of the locally affected poor.

In *Prafulla Samantray v Union of India*,[94] (popularly known as the POSCO case) the issue before the NGT was opposition to the proposed POSCO project involving the construction of an integrated steel plant with a service sea port at Paradip in the Jagatsinhpur district of the state of Orissa. The government of Orissa agreed to facilitate the project and assist the POSCO multinational steel company, based in South Korea, to obtain a no objection certificate and environmental clearances in the minimum time. The appellant, a social and environmental activist, challenged the proposed project on the ground that the manner in which the appraisal was undertaken – starting from the preparation of the environmental impact assessment report to conducting a public hearing to examination by the expert appraisal committee of the Ministry of Environment and Forests (MoEF) – was not in accordance with the environmental legislation. Further, the appellant contended that the long-standing forest dwelling communities, including betel leaf farmers, fisher folk and *dalits*, anticipated dislocation and loss of livelihood. There would also be an adverse impact on the environment and on agricultural and water resources. It was estimated that nearly 22,000 people would be affected by the proposed project. The people to be displaced were mostly agricultural and fishermen families (about 700 families). As POSCO was to be located on forest, land clearance under the Forest Conservation Act was necessary. The appellant alleged that the MoEF failed to consider the environmental and social implications of such a large project and relied mainly on the assurances given by POSCO. The MoEF and POSCO, on the other hand, denied the allegations made by the appellant and contended that the prescribed procedure for environmental clearances was followed in both letter and spirit.

The NGT allowed the appeal and observed:

---

[94] Judgment dated March 30 2012.

> we have kept in mind the need for industrial development, employment opportunities created by such projects that involve huge foreign investment, but at the same time we are conscious that any development should be within the parameters of environmental and ecological concerns and satisfying the principles of sustainable development and precautionary measures. A close scrutiny of the entire scheme of the process of issuing [a] final order in the light of the facts and material placed on record before NGT revealed that a project of this magnitude particularly in partnership with a foreign country has been dealt with casually, without there being any comprehensive scientific data regarding the possible environmental impacts. No meticulous scientific study was made on each and every aspect of the matter[,] leaving lingering and threatening environmental and ecological doubts unanswered. Further, the rights of the affected people were to be recognized before any forest land was acquired and resettlement and rehabilitation benefits worked out.[95]

Accordingly, the NGT suspended the approval erroneously granted to POSCO and directed the MoEF to conduct a fresh review in accordance with the law. The clearance granted was set aside as it was 'arbitrary and illegal' and 'vitiated in the eyes of law'.[96]

The suspension of the environmental clearance was welcomed by the affected people. Prafulla Samantray commented in a press note after the judgment that 'the recent verdict of the NGT shows that the people were right in their opposition and they will continue their struggle against any unjust or immoral imposition on them which would destabilize their life and livelihood'.[97] It is submitted that the NGT's verdict marked a victory for the human rights agenda.

Subsequently, the MoEF reviewed the matter and separated the project components into two parts – a steel plant and a port project. On 7 January 2014, the MoEF approved the building of the steel plant but subject to the condition that POSCO underwrites and performs 'social commitments' which will raise the project cost from US$600 million to US$12.6 billion.[98] The revalidated clearance was challenged before the NGT on 21 April 2014. The environmental clearance for the port is still pending. The forest clearance to acquire land has not been granted by the

---

95 Ibid para 7.

96 http://www.minesandcommunities.org/article.php?a=11597 (accessed 18.9.2014).

97 Press note dated March 31 2012, available at http://www.minesandcommunities.org/article.php?a=11597 (accessed 17.2.2014).

98 K Sabhav, 'Posco's Environmental Clearance Revalidated' (2014) *Down to Earth* January; http://zeenews.india.com/business/news/companies/posco-gets-environmental-clearance-for-rs-52k-odisha-project_92384.html (accessed 19.6.2014).

state, thereby moth-balling the POSCO project, which has remained embroiled in controversies concerning environmental clearances, land acquisition and displacement of forest dwelling communities since its announcement in 2007.

For the past eight years, the affected people, who are physically and culturally dependent on their environment, have faced violation of their human rights as a result of having to live their lives under siege – a price paid for voicing their opposition to POSCO and the state. An independent report, published by a Human Rights NGO in 2013, entitled 'The Price of Steel: Human Rights and Forced Evictions in the POSCO-India Project'[99] tells the story of a human rights crisis induced by mega-development projects that would lead to the forced migration of many thousands of poor people. The report states that:

> India's attempts to forcibly evict these communities stand in violation of international and domestic law. The Government of India must end human rights abuses tied to its project with the South Korean steel giant POSCO, and must immediately cease illegal seizure of land which threatens to forcibly displace as many as 22,000 people in the state of Odisha. Should the project move forward, entire villages will be decimated, livelihoods will be destroyed, and families will be rendered homeless, all in the name of 'development.' Entire communities in the project-affected area have been living under siege and have suffered clear violations of their rights to security of person and freedom of movement; their rights to be free from arbitrary arrest and detention; and their right to be free from discrimination – particularly on the basis of political or other opinions. Living under siege has also resulted in significant disruptions to many villagers' ability to access health care, schools, markets, and crops, undermining their rights to health, education, work, and food. On the environmental front, the construction of the proposed plant and port threatens the area's unique biodiversity. Many species of animals live and breed in the area's waters, lush estuaries, and dense forest. Among them is the endangered olive ridley turtle, which is protected under Indian law. The site of the proposed captive port includes beaches which serve as a nesting site for over 100,000 olive ridley turtles every year, one of only three such sites worldwide and the only site in the Eastern Hemisphere. Ecologically important horse shoe crabs also depend on the sediment along the coast in the port area, and fish and shrimps use the estuaries and coastal waters as breeding grounds. There also exist well-founded concerns regarding air pollution and the diversion of water resources.[100]

---

[99] International Human Rights Clinic, ESCR-Net, *The Price of Steel: Human Rights and Forced Evictions in the POSCO-India Project* (New York, NYU School of Law 2013).

[100] Ibid 1–4 and 15.

Access to justice through the NGT has revived hope for the poor, although the POSCO development outcome is yet to be settled. The Tribunal has sought to ensure that human rights and environmental considerations are reconciled in order to ensure sustainability, social equity and an inclusive interpretation of 'development'. In the meantime, minute by minute, the potentially affected people remain in high anxiety, unsure of their families' future, residences, land, incomes and social relationships.

Another illustration of competing interests is the construction of dams. Water is an essential, life sustaining element that places responsibility upon the State to ensure its appropriate and equitable collection, storage, availability and usage. India is monsoon dependent for its water and the irregular supply of water is under increasing, competitive pressure from industrialisation, new farming processes and a growing population. There is a continuing debate concerning the effectiveness and environmental impact of large as opposed to multiple small dams. Large dams involve submergence or destruction of forests and have significant social consequences, including large-scale rehabilitation and resettlement, family and community disruption and employment issues for relocated tribal people. This debate is well illustrated by the Sardar Sarovar Dam, which is based in Maharashtra.[101] The importance of the caretaker role of the State and of relevant government agencies in approving appropriate development projects after assessing the technical data, social and environmental impact studies and compliance with environmental legal requirements cannot be over-emphasised. It is at this point, however, that crucial failures occur.

The NGT became involved in a mega dam dispute when the required legal procedures prior to construction of the Mapithel dam were ignored by the MoEF and by the state of Manipur. The completed dam will submerge 778 hectares of agricultural land and 595 hectares of forest land, resulting in the displacement of many thousands of people who are mainly tribal. Under the Forest Rights Act 2006, certain procedures must be followed before permission can be granted to use forest land for non-forest purposes. The Act requires that affected people must be consulted and must approve the change of use. Since almost 80 per cent of the affected villagers are dependent on agriculture and on forest produce it is perhaps unsurprising that state officials attempted to

[101] P Cullet, *The Sardar Sarovar Dam Project* (Aldershot, Ashgate 2007) 1–3; *The Economist*, 'The Ups and Downs of Dams' May 2010 (Special Report) available at http://www.economist.com/node/16136280 (accessed 18.3.2014).

circumvent the legislation and to ignore the rights of the local people. The villagers took their grievances to the NGT in the case of *Themrei Tuithung v State of Manipur*,[102] arguing that the dam project did not have the necessary clearance from the Union Ministry of Environment and Forests. More generally, it was stated that the neo-liberal policies of Manipur state, which had attracted significant inward infra-structure investment, were exploiting natural resources at the expense of local people. The result was growing land alienation, community displacement and various human rights violations. In November 2013, the NGT suspended construction of the dam on the grounds of non-compliance with the relevant environmental protection regulations and in the public interest.[103]

In *Lower Painganga Dharan Virodhi v State of Maharashtra*,[104] the NGT was faced with the issue of deciding on dam construction approval in the light of its social, economic and environmental consequences. The applicants approached the NGT to quash the environmental clearance issued for the construction of an irrigation project (dam) in order to protect the ecology and interests of tribal farmers and villagers. According to the applicants, the proposed project would adversely affect 90,000 hectares of land, including forest land. Nearly 2 lakh trees (200,000 trees) were likely to be felled causing irreversible damage to the environment. Furthermore, the project would displace 7,102 families consisting of 35,388 tribal people whose families have lived in the area for generations. It was alleged that the environmental clearance was illegal, being without application of mind and devoid of reason and was obtained illegally by the respondents. The procedural requirements of a public hearing – due participation of affected public members; publication of notice of the hearing in at least two newspapers, one thereof being in vernacular language; and a copy of the environmental law assessment report made available to public – were not followed. The land on which the tribal farmers survived was being acquired without permission from village councils, thereby flouting forest related laws.[105] The irrigation project was opposed by the inhabitants of 95 villages and their opinion was not taken into account.

The respondents contended that the environmental clearance was lawfully obtained. They submitted a special study conducted by an expert

---

[102] Original Application Number 167/2013 dated August 5 2013.
[103] Order dated November 12 2013.
[104] Judgment dated March 10 2014.
[105] The Forest (Conservation) Act 1980 and Schedule IV of the Environmental Impact Assessment Notification 1994.

agency that had evaluated the feasibility of the irrigation project. The respondents provided a plan for the rehabilitation and resettlement of the affected tribal people to the Ministry of Tribal Affairs, Government of India, in order to ensure that the affected people were compensated and rehabilitated.

The NGT, after going through the voluminous records and documents, was of the opinion that the irrigation project satisfied the principle of sustainable development and that the environmental safeguards proposed by the project proponent and stipulated by the regulatory authorities was adequately incorporated. The Tribunal stated:

> [T]he dams are a large infrastructure and have a high potential for development[;] they can balance hydrological variability by storing water for all sectors of the society and serve for controlling the floods. Irrigation project envisages benefits to the tribals, farmers of socially and economically backward area of Vidarbha and aims to generate employment in that area. Nobody will deny that a major irrigation project is likely to give booster dose to the economy of the region. Availability of irrigation facilities in the area will help cultivators to minimize or curtail dependency on annual rainfall, which is many times unpredictable. The right to development includes, of course, protection of fundamental human rights. Thus, if the irrigation project is required for [the] larger benefit of the society, then it must be ensured that the Project Affected Families are given [a] justifiable rehabilitation package. Rehabilitation is not only about providing basic amenities, food and shelter. It encompasses support to restore means of livelihood. Rehabilitation of Project Affected Persons (PAP) is logical fallout of Article 21. They could not be left at the mercy of others. The project proponent has to take care of the future needs of PAP. Prior rehabilitation will create a sense of confidence among the project affected family members and they will be in a better position to support their life on their own, at the same time, when the project would become functional.[106]

Despite the recognition by the NGT of the rights of poor people and the need to allow development subject to environmental regulations, a shocking report by the Comptroller and Auditor General (CAG) on the Management of Irrigation Projects in Maharashtra, June 2014,[107] condemns the state government's repeated failure to undertake its statutory environmental protection duties and social responsibilities in regard to

---

[106] Ibid paras 41 and 44.

[107] South Asia Network on Dams, Rivers and People, 'Rampant Environmental Violation of Maharashtra Water Resource Department' June 21 2014, available at http://sandrp.wordpress.com/2014/06/20/press-release21-06-14-rampant-environmental-violations-of-maharashtra-water-resource-department/ (accessed 22.6.2014).

water management. The CAG report states that as many as 249 projects of the Vidharbha Irrigation Development Corporation (VIDC) commenced without receiving the necessary Environmental Clearance (EC) from the Central Ministry of Environment and Forests or the State Environment Department.

In summary, it can be seen that the NGT's decisions seek to strike a delicate balance involving human rights and the environment, particularly in the context of resource exploitation and public utility projects. However, it appears that the efforts of NGT to challenge the process that undermines green concerns and the rights of the people in order to facilitate unbridled growth is being hampered by the government's economic policies. Increasing political pressure is being exerted to promote economic development via public and private investment at the cost of the environment. The government is planning to amend the forest laws[108] to increase economic activities in states,[109] which hold some of India's most productive forests and also the majority of the tribal population. The government also seeks to dilute the Forest Rights Act[110] that requires the consent of the local tribal population concerning change of use of forest land, particularly mining for minerals. Prospecting in forests previously involved the approval of the gram sabhas (village councils) certifying that tribal rights had been settled and that they had consented to projects. Now the district administrations will have this responsibility thereby excluding project affected tribal people from the final decision-making process.

It is hoped that the NGT will hold fast to its basic and established jurisprudence in the face of these new economic policies.

---

[108] The Scheduled Tribes and Other Traditional Forest Dwellers [Recognition of Forest Rights Act] 2006 and the Forest Conservation Act 1980. http://articles.economictimes.indiatimes.com/2014-08-28/news/53324639_1_environment-ministry-prakash-javadekar-irrigation-projects (accessed 18.9.2014).

[109] The following states, Chhattisgarh, Jharkhand, Odisha, Maharashtra, Andhra Pradesh and West Bengal have active armed insurrection from Maoists. The government is focusing on economic development, particularly public projects to tackle insurgency. In addition there is an increased armed response towards Maoists.

[110] The Scheduled Tribes and Other Traditional Forest Dwellers [Recognition of Forest Rights Act] 2006.

## 4. CONCLUSION

The constitutionally-based judicial journey from public interest litigation to the National Green Tribunal continues. The successes and short-comings of public interest litigation and of environmental protection must be contextualised within the wider socio-economic world of Indian politics. Accordingly, to view this pathway as solely legal offers a distorted and limited understanding. The size and complexity of 'India' ensures that legal decision-making is subject to many pressures. For example, the chronic issues of administrative inefficiency, maladministration and endemic corruption are deep rooted and continue to frustrate the implementation of judicial orders. Prime Minister Modi's commitment to stimulate economic growth as his top target includes fast tracking environmental clearances to support industrial construction, both in the public and private sectors. Additionally, the government is considering reviewing and reducing the statutory powers of the NGT. It is suggested that this is a consequence of the NGT's decisions, which highlight the failure of state authorities and environment ministry officials to perform their statutory duties.[111]

On the other hand, the historic importance of the Supreme Court; its creation of the procedural tool of public interest litigation, which provides access to justice for poor people; and its willingness to speak out for the 'under-dog' in environmental cases is a matter of public record. Today, it is the turn of the National Green Tribunal to assume the principal position in adjudicating environmental issues and to build upon the environmental jurisprudence of the Supreme Court. The Tribunal's case law indicates a willingness to support development, but only if subjected to the strict application of the NGT Act 2010, its constitutional mandate and the relevant environmental regulations. The analysis offered in this chapter demonstrates that the Tribunal's decisions continue to reflect an understanding and appreciation of, and a commitment to harmonising, human right principles of a right to a healthy environment and also to sustainable development.

---

[111] http://www.business-standard.com/article/economy-policy/government-planning-to-clip-national-green-tribunal-s-wings-114080600015_1.html?_ga=1.31411512.620261427.1411058219 (accessed 18.9.2014).

# 6. Re-imagining adjudication: human rights courts and the environment

**Evadne Grant**

## 1. INTRODUCTION

The aim of the 2012 GNHRE Oñati International Workshop, 'Human Rights and the Environment: In Search of a New Relationship', was to explore ways of re-imagining the relationship between human rights and the environment.[1] In the course of thought-provoking conversations around papers exploring aspects of the philosophical, legal and practical dimensions of this relationship,[2] one of the most pervasive themes that emerged was the importance of re-establishing vital connections, not only between human rights and the environment, but also between rights, and between individuals, communities and rights.[3]

As Cormac Cullinan reflects in *Wild Law: A Manifesto for Earth Justice*:

> The human societies that presently dominate the world govern on the basis of a false understanding of the universe. The core falsehood is that we humans are separate from our environment and that we can flourish even as the health of Earth deteriorates. … The governance structures, legal philosophies (jurisprudence), and laws established by many societies reflect and entrench the illusion of separation and independence. This encourages and legitimises

[1] A Grear, 'Human Rights and the Environment: In Search of a New Relationship: Editor's Introduction' (2013) 3 *Oñati Socio-Legal Series*, available at SSRN http://ssrn.com/abstract=2247824 (accessed 2.11.2014).

[2] See A Grear, 'Towards a New Horizon: In Search of a Renewing Socio-Juridical Imaginary' (2013) 2 *Oñati Socio-Legal Series*, available at SSRN http://ssrn.com/abstract=2247857 (accessed 2.11.2014).

[3] E Grant, L Kotze and KL Morrow, 'Human Rights and the Environment: In Search of a New Relationship. Synergies and Common Themes' (2013) 3 *Oñati Socio-Legal Series*, available at SSRN http://ssrn.com/abstract=2221302 (accessed 2.11.2014).

> environmentally and socially destructive behaviour and hinders the development of more appropriate forms of social organisation …[4]

The significance of the very close connection between human rights and the environment is increasingly being recognised, within international human rights law as well as environmental law.[5] The 2012 Preliminary Report of John Knox, Independent Expert on Human Rights and the Environment, provides a clear example of the recognition that '[h]uman rights and environmental protection are inherently interdependent'.[6] An understanding of the interconnectedness of human rights and the environment is also increasingly reflected in the practice of human rights courts and tribunals, at national, international and regional levels,[7] in their response to multiplying complaints raising environmental issues – from claims of rights violations due to the destructive impact of mining and oil extraction[8] to environmental degradation of the lands on which indigenous peoples depend.[9]

But while there is increasing understanding that many, if not all, of the rights protected in international human rights law have an environmental dimension, the extent to which human rights courts are able to incorporate environmental protection into human rights cases is disputed. Critics argue, for example, that a human rights approach to environmental protection is severely limited by conceptual and procedural constraints inherent to human rights law, ensuring that attention remains focused on

---

[4] C Cullinan, *Wild Law: A Manifesto for Earth Justice* (2nd edn, Green Books 2003) 44.

[5] See for example, *Declaration of the UN Conference on the Human Environment*, UN Doc A/CONF.48/14/Rev 1, 3 (16 June 1972) (Stockholm Declaration) Principle 1; S Chuffart and JE Viñuales, 'From the Other Shore: Economic, Social and Cultural Rights from an International Environmental Law Perspective' in E Riedel, G Giacca and D Golay (eds), *Economic, Social and Cultural Rights in International Law: Contemporary Issues and Challenges* (Oxford University Press 2014).

[6] *Report of the Independent Expert on the issue of human rights obligations relating to the enjoyment of a safe, clean healthy and sustainable environment: Preliminary report*, A/HRC/22/43, 24 December 2012, para 10.

[7] D Shelton, 'Developing Substantive Environmental Rights' (2010) 1 *JHRE* 89, 90.

[8] For example, *The Social and Economic Rights Action Centre and the Centre for Economic and Social Rights (SERAC) v Nigeria*, Communication 155/96, (2001).

[9] For example, *Maya Indigenous Communities of the Toledo District (Belize)*, Case 12.053 Report No 40/04, Inter-American Commission on Human Rights, (12 October 2004).

human well-being and confining the legal protection of the environment to situations in which the environment is useful to humans.[10] Such constraints, it is argued, arise from a failure to recognise the importance of, and to engage with, the myriad ways in which individuals, communities and the environment are interconnected. For example, it is often argued that human rights adjudication is inherently individualistic, and that consideration of the environmental dimension of rights will therefore necessarily be limited to the extent to which environmental harms affect particular individual claimants. Accordingly, it is concluded by some that the essential link between the individual and the collective interests of whole communities affected by environmental degradation and indeed all humans, in maintaining a healthy and sustainable environment is often ignored.[11] It is also argued that human rights adjudication tends to focus on isolated rights rather than on the relationship between rights, in spite of the endorsement of the idea of human rights as 'indivisible, interdependent and interrelated' by the international community in the 1993 Vienna Declaration.[12] Accordingly, the fact that environmental damage usually does not impact just one right, but multiple rights, is often disregarded – it is thought – by human rights courts, which seek to confine their judgments to specific individual rights rather than addressing the widespread impact of environmental harm on multiple human rights.

The problem of disconnection manifests itself in a number of specific constraints in how environmental concerns are dealt with in human rights adjudication. For example, rules of standing which determine who can bring claims in human rights courts often limit opportunities for environmentally destructive actions to be challenged via human rights law.

---

[10] C Gearty, 'Do Human Rights Help or Hinder Environmental Protection?' (2010) 1 *JHRE* 7, 7–9; S Coyle and K Morrow, *The Philosophical Foundations of Environmental Law: Property, Rights and Nature* (Hart 2004) 211; C Gianolla, 'Human Rights and Nature: Intercultural Perspectives and International Aspirations' (2013) 4 *JHRE* 58, 63.

[11] F Francioni, 'International Human Rights in an Environmental Horizon' (2010) 21 *EJIL* 41, 42, 44; R Pavoni, 'Environmental Jurisprudence of the European and Inter-American Courts of Human Rights: Comparative Insights' in B Boer (ed.), *The Environmental Dimension of Human Rights* (Oxford University Press 2015) 69.

[12] World Conference on Human Rights, 'Vienna Declaration and Programme of Action' (1993) UN Doc A/CONF 157/23, para 5. See E Grant, 'Accountability for Human Rights Abuses: Taking the Universality, Indivisibility, Interdependence and Interrelatedness of Human Rights Seriously' (2007) 32 *SAYIL* 158.

Human rights courts also often refuse to consider the impact of environmental harm on whole communities, only providing opportunities for vindication of particular rights of particular individuals. The formulation of rights in many international instruments reinforces a view of rights as being disconnected: rights are defined, more often than not, in isolation from each other and from the environmental context. The adoption of separate treaties relating to civil and political rights and to social and economic rights emphasises and reinforces the disconnection. And while considerable progress has been made in developing theoretical justification of environmental protection, particularly as it relates to social and economic rights,[13] the jurisdiction of human rights courts over social and economic rights is often restricted so that the potential for social and economic rights adjudication to support and advance environmental protection is constrained.

There is undoubtedly much truth in this broadly critical account of human rights individualism and disconnectedness. Yet, while the discussion above might suggest a rather negative outlook for the use of human rights law in providing environmental protection, the reality – for all the undoubted purchase of the critiques offered of the weaknesses of human rights – is more encouraging.

In spite of significant constraints, the past three or four decades have seen the development of a substantial body of law in which the interconnectedness of human rights and the environment has been recognised and elaborated. In this regard, a number of different threads of what can be referred to as 'environmental human rights' can be identified – in particular developments around the understanding that a healthy and sustainable environment is a precondition for the enjoyment of all human rights and the recognition that procedural rights, such as the right to information and participation, are important to environmental protection and are inherent in a number of human rights when applied in an environmental context.[14]

The development of environmental protection via recognition that a sustainable environment is a precondition for adequate protection of all

[13] See OUNHCHR, *Mapping Human Rights Obligations Relating to the Enjoyment of a Safe, Clean, Healthy and Sustainable Environment: Individual Report on the International Covenant on Economic, Social and Cultural Rights*, December 2013 available at http://ieenvironment.org/2014/03/06/2014-mapping-reports/ (accessed 12.11.2014).

[14] Shelton (n 7) 90; K Hulme, 'International Environmental Law and Human Rights' in S Sheeran and N Rodley, *Routledge Handbook of International Human Rights Law* (Routledge 2013) 285, 289.

human rights is being taken forward by a range of human rights institutions – at national, regional and international levels. The Committee on Economic Social and Cultural Rights (CESCR), for example, has made a significant contribution to the recognition of the importance of a clean, healthy and sustainable environment for the protection of a variety of social and economic rights by its interpretation of the rights protected under the International Covenant on Economic, Social and Cultural Rights (ICESCR).[15] Human rights courts have provided a particularly important avenue for the expansion of environmental protection in human rights law, especially in the absence of a compulsory dispute resolution mechanism in international environmental law. Indeed, as the analysis offered here will reveal, the process of adjudication is of fundamental importance in the development of environmental principles and the recognition of international environmental rules and norms.[16] There have also been important developments in relation to the recognition of a substantive environmental right: although, as will be discussed below, the African Charter on Human and Peoples' Rights (ACHPR)[17] is the only regional human rights treaty that specifically recognises a right to a clean and sustainable environment, a significant number of national constitutions now include substantive environmental rights.[18]

Focusing on key cases decided by regional human rights institutions, this chapter seeks to interrogate the jurisprudence in order to assess the extent to which – and the ways in which – human rights courts recognise the interconnectedness of human rights and the environment at multiple levels, and to consider how existing practice might assist in re-imagining the relationship between human rights and the environment in order better to protect the environment.

## 2. REGIONAL HUMAN RIGHTS FRAMEWORKS

As already noted, among the three regional human rights systems, only the ACHPR[19] expressly recognises a substantive environmental human

---

[15] OUNHCHR (n 13).

[16] T Stephens, *International Courts and Environmental Protection* (Cambridge University Press 2009) 12.

[17] African Charter on Human and Peoples' Rights (adopted 27 June 1981, entered into force 21 October 1989) OAU Doc CAB/LEG/67/3 Rev 5; 1520 UNTS 217; (1982) 21 ILM 58 (ACHPR).

[18] See DR Boyd, *The Environmental Rights Revolution* (UBC Press 2012).

[19] ACHPR (n 17).

right, expressed in the following terms: 'All peoples have the right to a general satisfactory environment favourable to their development'.[20] The ACHPR is also the only regional human rights treaty that clearly recognises the interconnectedness and interdependence of all human rights, stating explicitly that 'the satisfaction of economic, social and cultural rights is a guarantee for the enjoyment of civil and political rights'.[21] This is reflected in the text of the ACHPR itself, which draws no distinction between civil and political rights and social and economic rights. Social and economic rights are thus not qualified by reference to 'progressive realisation' and 'available resources', as is the case in other international instruments which provide protection for economic, social and cultural rights.[22]

The European Convention on Human Rights (ECHR)[23] is less attuned to the interconnectedness of human rights and the environment. The Convention itself provides for civil and political rights only, while social and economic rights are provided for in a separate treaty – the European Social Charter (ESC) with its own implementation mechanism.[24] There is no provision for environmental rights in either treaty. Accession to the ECHR is a condition of Council of Europe membership, but this is not required for the ESC, reinforcing the disconnection between civil and political rights and social and cultural rights. The jurisdiction of the European Court of Human Rights (ECtHR) is limited to the ECHR. The Inter-American human rights system similarly clearly separates civil and political rights and social and economic rights. The American Convention on Human Rights (ACHR) refers to social and economic rights in one

---

20 Ibid Article 24. The San Salvador Protocol to the American Convention on Human Rights also provides for a specific environmental right, but its application is limited. See Additional Protocol to the American Convention on Human Rights in the Area of Economic Social and Cultural Rights (entered into force 16 November 1999) OAS Treaty Series 69; 28 *ILM* 156 (1989).

21 ACHPR (n 17), Preamble.

22 See for example, International Covenant on Economic, Social and Cultural Rights (adopted 16 December 1966, entered into force 3 January 1976) 993 UNTS 3 (ICESCR) Article 11.

23 European Convention for the Protection of Human Rights and Fundamental Freedoms (as amended) (adopted 4 November 1950, entered into force 3 September 1953) ETS 5; 213 UNTS 222 (ECHR).

24 European Social Charter (adopted 18 October 1961, entered into force 26 February 1965) ETS 35; 529 UNTS 89 (ESC).

provision only,[25] while the San Salvador Protocol to the American Convention[26] makes separate provision for social and economic rights including a right to a healthy environment.[27] The Inter-American Commission of Human Rights (IACommHR) and Inter-American Court of Human Rights (IACtHR) have limited jurisdiction in relation to the rights contained in the Protocol.[28] The clear message that both the ECHR and the ACHR frameworks convey is that social and economic rights are less important than civil and political rights and thus enjoy less protection. This approach has implications for protection of the environment since a clean, healthy and sustainable environment is most closely associated with social and economic rights such as the right to food, water and an adequate standard of living.

The view that civil and political rights are distinct from and unconnected to social, economic and cultural rights is clearly demonstrated by the brief overview of the principal regional human rights instruments. The limited extent to which specific provision is made for protection of the environment is also evident. However, the exact terms of any international human rights instrument are merely a starting point for the development of rights protection. It is through interpretation, elaboration and application by human rights institutions that the meaning and scope of human rights are established. The extent to which environmental protection can be realised via human rights law is thus crucially dependent on the way in which human rights are interpreted by human rights bodies, especially human rights courts.

## 3. CONNECTING RIGHTS AND COMMUNITIES

As noted above, human rights adjudication often limits consideration of the environmental dimension to the effect thereof on a single individual and in relation to a single disconnected right. Such an approach disregards the potential impact of environmental harm on multiple people

---

[25] American Convention on Human Rights (adopted 21 November 1969, entered into force 18 July 1978) OAS Treaty Series Rights 36; 1144 UNTS 123; 9 ILM 99 (1969) Article 26.

[26] San Salvador Protocol (n 20).

[27] Ibid Article 11.

[28] See OR Ruiz-Chiriboga, 'The American Convention and the Protocol of San Salvador: Two Intertwined Treaties – Non-Enforceability of Economic, Social and Cultural Rights in the Inter-American System' (2013) 31 *NQHR* 159.

and multiple rights and on the collective interest in maintaining a healthy and sustainable environment.[29]

One of the key ways in which such a narrow focus manifests itself in the practice of courts is, as already observed, through restrictions placed by human rights instruments on who may bring complaints of violations before human rights courts. In terms of Article 34 of the ECHR, for example, in order for an applicant to bring a case to the ECtHR, he or she must be a 'victim' of a violation of one of the rights protected under the Convention. The 'victim' requirement means that an applicant must be 'personally affected by an alleged violation of a Convention right' according to the ECtHR.[30] One of the implications of this is that the ECHR system makes no provision for public interest claims for the protection of the environment as a common good,[31] and the ECtHR has emphasised in a number of cases that the ECHR does not provide protection of the environment as such[32] and that challenges 'in abstracto', which do not relate to a breach of specific rights under the convention, are not permitted.[33]

In spite of its narrow approach to standing, the ECtHR has made significant progress in broadening the interpretation of the rights protected under the ECHR in order to provide protection for the environment on the basis that environmental harm may significantly impact the enjoyment of a range of rights protected under the ECHR. The ECtHR has acknowledged that rights, such as the right to life (Article 2), the right to private and family life (Article 8) and the right to a fair trial (Article 6) have an environmental dimension and that environmental damage may lead to violation of those rights. The jurisprudence is most developed in relation to the right to private and family life, where the ECtHR has recognised that excessive noise,[34] smells,[35] and other types of

---

[29] Francioni (n 11) 44.

[30] *Karner v Austria*, Application No 40016/98, 24 July 2003, para 25.

[31] *Ilhan v Turkey*, Application No 22277/93, 27 June 2000, paras 52–53; see R Pavoni, 'Public Interest Environmental Litigation and the European Court of Human Rights: No Love at First Sight' in F Lenzerini and AF Vrdoljak (eds), *International Law for Common Goods: Normative Perspectives on Human Rights, Culture and Nature* (Hart 2014) 331, 333.

[32] Council of Europe, *Manual on Human Rights and the Environment* (Council of Europe Publishing 2012) 7; Shelton (n 7) 106.

[33] *Karner v Austria* (n 30) para 24.

[34] *Hatton and Others v the United Kingdom*, Application No 36022/97, 8 July 2003.

[35] *López Ostra v Spain*, Application No 16798/90, 9 December 1994.

serious environmental pollution[36] that adversely affect the well-being of individuals and prevent them from enjoying their homes, may violate the right to private and family life.[37] While not precisely articulated, the Article 8 jurisprudence demonstrates an understanding on the part of the ECtHR of the interconnectedness, not only of human rights and the environment, but also between rights, particularly between the right to private and family life, the right to health and the right to a healthy environment.

However, because of the overriding focus on individual rights, the Article 8 jurisprudence remains limited in its recognition of the importance of a healthy and sustainable environment to whole communities. Thus, the environmental dimension of Article 8 will only play a role if it can be shown that that environmental harm had a direct impact on a particular individual in relation to that particular person's rights; the impact on whole communities is not relevant.[38] How this plays out in practice can be seen in a range of cases.

The central question for the ECtHR in deciding cases under Article 8 is whether a direct causal link between environmental harm and the ability of the individual claimant to enjoy his or her home can be established. In order to engage Article 8, damage to the environment must be shown to impact the health or well-being of the claimant to such an extent that it affects the enjoyment of his or her home.[39] In *Kyrtatos v Greece*,[40] for example, the Court rejected a claim under Article 8 since, in the view of the Court, it had not been shown that the environmental damage at issue had a direct impact on the well-being of the applicants and their ability to enjoy their home. The applicants complained that the destruction of a wetland close to their property, which was a natural habitat for wildlife, affected their enjoyment of their home and that the Greek authorities had by their actions and inaction, failed to protect their rights under Article 8. The applicants accepted that urban development was necessary for economic reasons, but argued that any interference with the environment should strike a balance between the competing interests of individuals

---

[36] *Guerra and Others v Italy*, Application No 116/1996/735/932, 19 February 1998.

[37] Council of Europe, *Manual on Human Rights and the Environment* (n 32) 45; D Harris, M O'Boyle, EP Bates and CM Buckely, *Harris, O'Boyle & Warbrick: Law of the European Convention on Human Rights* (3rd edn, Oxford University Press 2014).

[38] Francioni (n 11) 50.

[39] Ibid.

[40] *Kyrtatos v Greece*, Application No 41666/98, 22 May 2003.

and the community interest in the environment. The ECtHR accepted that it was not necessary for the applicants to show that their health had been affected by environmental factors in order for their claim under Article 8 to succeed, since a breach of the right to private and family life could arise as a result of environmental pollution which affected the well-being of individuals, thus preventing them from enjoying their homes.[41] However, the Court held that the applicants 'had not brought forward any convincing arguments showing that the alleged damage to the birds and other protected species living in the swamp was of such a nature as to directly affect their own rights under Article 8'.[42] *Kyrtatos v Greece* thus demonstrates the limitations of an approach which focuses exclusively on the impact of environmental harm on individual applicants and which excludes consideration of the wider effect of destruction of the environment.

In contrast to *Kyrtatos*, however, the case of *Taşkin v Turkey*[43] demonstrates the extension of environmental protection, albeit in a different direction. The applicants in *Taşkin* raised concerns about the risks to the environment, especially the risk of groundwater contamination and damage to wildlife, posed by a process using sodium cyanide at a local gold mine. The applicants argued that the risk posed could potentially affect their rights under Article 8 ECHR. An environmental impact report had been prepared for the Turkish government, which had nonetheless granted permission for mining activities to go ahead. The government subsequently ignored a judgment of the Supreme Administrative Court annulling the decision to allow the mining activity, finding that risk to the environment and human health highlighted by the report was very high. The Turkish government argued that the risk to the applicants was purely hypothetical and that in the absence of a 'serious and imminent risk',[44] Article 8 was inapplicable. The ECtHR disagreed, holding that Article 8 applied if an environmental impact assessment had established a serious risk that was *likely* to affect the applicants in such a way as to affect their private and family life. In the view of the court, this conclusion is analogous to the applicability of Article 8 to situations in which environmental pollution affects an individual's well-being and thus affects their enjoyment of their homes, even if there is no serious danger

---

41 Ibid, para 52 citing *López Ostra v Spain* (n 35).

42 *Kyrtatos* (n 40), para 53. See also *Tătar v Romania* Application No 657021/01, 27 January 2009.

43 *Taşkin and Others v Turkey,* Application No 46117/99, 10 November 2004.

44 Ibid para 107.

to their health.[45] *Taşkin* represents an important development, allowing the ECtHR to consider risks to the environment rather than having to wait until actual damage occurs. However, the potential impact on an individual applicant remains the focus of investigation – the potential effect of pollution or other environmental dangers on whole communities and habitats is only relevant if it is shown to impact the well-being of a particular individual who applies to the court.

In direct contrast to the ECtHR approach that disconnects individuals from communities, individualises rights and disregards the large-scale impact of environmental destruction, cases decided by both the African and Inter-American human rights institutions indicate a far greater willingness to develop the environmental dimension of human rights, recognising that environmental integrity is essential for the realisation of human rights. Extension of protection by recognising group claims, for example – an approach typical of the African and Inter-American institutions – is an important step in the development of a broader approach to environmental human rights that recognises that most environmental claims concern whole communities rather than isolated individuals.

Standing to bring claims under the American Convention on Human Rights (ACHR) is not limited to applicants who satisfy a 'victim' requirement as under the ECHR.[46] While it is still necessary for injured parties to be identified, relatives or NGOs can bring cases to the IACommHR, which can refer matters on to the Inter-American Court of Human Rights (IACtHR).[47] Although this substantially broadens access to the IACommHR, it should be noted that violation of the rights of particular, identifiable human victims is still required in order to engage the IACommHR and IACtHR and that public interest claims are

---

[45] *López Ostra v Spain* (n 35); see Shelton (n 7) 115.

[46] Article 44 of the ACHR states that '[a]ny person or group of persons, or any nongovernmental entity legally recognized in one or more member States of the Organization, may lodge petitions with the commission containing denunciations or complaints of violations of this Convention by a State Party'.

[47] The IACommHR is empowered to consider cases brought by individuals, groups or NGOs (Article 44 ACHR) and may submit cases to the IACtHR where it finds a violation and a friendly settlement cannot be reached (Article 51 ACHR). The IACtHR has contentious jurisdiction over cases concerning alleged violation of the ACHR but the states concerned must recognise its competence to hear such cases and only the Commission or States Parties can submit cases (Articles 61 and 62 ACHR). Cases against states who are not party to the ACHR or who do not recognise the jurisdiction of the IACtHR can only be considered by the Commission.

excluded.[48] For example, in *Metropolitan Nature Reserve (Panama)* the IACommHR held that a petition brought on behalf of the citizens of Panama challenging the construction of a road through a nature reserve was inadmissible 'since it concerns abstract victims represented in an *actio popularis* rather than specifically identified and defined individuals'.[49] At the same time, however, the IACommHR confirmed the admissibility of claims brought by or on behalf of 'groups of victims when the group itself was specifically defined, and when the respective rights of identifiable individual members were directly impaired by the situation giving rise to a stated complaint'.[50] In practice, this approach has been applied specifically in cases involving indigenous communities[51] and it is in this context that much of the environmental jurisprudence of the IACommHR and IACtHR has been developed.[52]

In *Maya Indigenous Communities of the Toledo District (Belize)*[53] a petition brought on behalf of the Maya people claimed violation of a range of rights by the State of Belize. The petitioners argued that the granting of logging and oil concessions by the State on land traditionally occupied and used by the Maya people had resulted in damage to the environment and had consequently violated their right to property and endangered their culture and future existence. The IACommHR noted that:

> [T]he organs of the inter-American human rights system have acknowledged that indigenous peoples enjoy a particular relationship with the lands and resources traditionally occupied and used by them, by which those lands

---

48 *International Responsibility for the Promulgation and Enforcement of Laws in Violation of the Convention (Articles 1 and 2 of the American Convention on Human Rights)*, Advisory Opinion OC-14/94, Inter-American Court of Human Rights Series A No 14 ( 9 December 1994) para 45; see DK Anton and DL Shelton, *Environmental Protection and Human Rights* (Cambridge University Press 2011) 514.

49 *Metropolitan Nature Reserve v Panama*, Case 11.533, Report No 88/03, Inter-American Commission on Human Rights (22 October 2003) para 34.

50 Ibid para 32.

51 See for example, *Saramaka People v Suriname* (Preliminary Objections, Merits, Reparations, and Costs) Inter-American Court of Human Rights Series C No 172 (28 November 2007).

52 See for example, *Mayagna (Sumo) Awas Tingni Community v Nicaragua* (Merits, Reparations and Costs) Inter-American Court of Human Rights Series C No 79 (31 August 2001) (logging concessions); *Maya Indigenous Communities of the Toledo District (Belize)*, Case 12.053, Report No 40/04, Inter-American Commission on Human Rights (12 October 2004) (logging and oil concessions).

53 *Maya Indigenous Communities of the Toledo District (Belize)*, ibid.

> and resources are considered to be owned and enjoyed by the indigenous community as a whole and according to which the use and enjoyment of the land and its resources are integral components of the physical and cultural survival of the indigenous communities and the effective realization of their human rights more broadly.[54]

The IACommHR concluded that the State had violated the right to property of the Maya people by its failure to recognise and protect the communal right to the lands traditionally occupied and used by them.[55]

The case of *Saramaka People v Suriname*[56] saw the recognition of communal rights to property of non-indigenous groups. The applicability of the right to communal property of indigenous people to the facts of the case was disputed, as the Saramaka people are not indigenous to the region they inhabit but are the descendants of slaves who had been brought to Suriname during the colonial period.[57] The IACtHR accepted that, although the Saramaka people were not indigenous, they were a distinct social, cultural and economic group who identify strongly with their ancestral territories and have their own customs and traditions.[58] The court therefore recognised the Saramaka people as a tribal community similar to indigenous people, holding that they were entitled to the same protection of their communal right to property as indigenous peoples.[59] The court went on to emphasise that the rights of indigenous communities to communal property[60] included ownership of the natural resources necessary to enable communities to continue their traditional way of life.[61] The close connection between the protection of indigenous land and environmental protection was highlighted by the Court, which held that in order to safeguard the right to property of communities, an independent environmental and social impact assessment must be carried out before permission is granted for resource exploitation.[62]

---

54 Ibid para 114.

55 Ibid para 151–53. The failure to consult with the Maya people was considered to be an aspect of the violation of the right to property by the IACommHR (discussed below).

56 *Saramaka People v Suriname* (n 51).

57 Ibid paras 78–80.

58 Ibid paras 79–80.

59 Ibid para 86.

60 Ibid paras 89, 122.

61 Ibid para 121.

62 Ibid para 129. Consultation, consent and benefit sharing were also required according to the IACommHR (see discussion of procedural rights below).

The more relaxed approach to standing taking by the inter-American institutions loosens the grip of individualism over human rights claims, despite the fact that the rejection of public interest claims by the IACtHR continues to limit access for environmental claims. Nonetheless, the recognition of group claims in relation to the right to property of indigenous and tribal peoples begins to open up other possibilities for the wider recognition of communal rights to a sustainable and healthy environment – with progressive potential implications, moreover, for the protection of indigenous environments.

The fact that the ACHPR – uniquely – incorporates a substantive environmental human right specifically formulated as a group right greatly facilitates access to the ACommHPR and ACtHPR in environmental matters.[63] As in the Inter-American system, there is no 'victim' requirement under the ACHPR[64] and specific provision is made for NGOs, including international NGOs, to bring matters before the Commission and Court.[65] But the African system goes further than the Inter-American system by not requiring applicants to show that they are directly affected by the alleged breach.[66] It is not even necessary for

---

[63] See SA Dersso, 'The Jurisprudence of the African Commission on Human and Peoples' Rights with Respect to Peoples' Rights' (2006) 6 *African Human Rights Law Journa*l 358.

[64] Article 56 ACHPR and Article 5(3) of the Protocol to the African Charter on Human and Peoples' Rights on the Establishment of an African Court on Human and Peoples' Rights (Protocol on the African Court) OAU/LEG/AFCHPR/PROT (III), adopted by the Assembly of Heads of State and Government, 34th Session, 8–10 June 1998; see F Viljoen, 'Admissibility under the African Charter' in M Evans and R Murray, *The African Charter on Human and Peoples' Rights* (Cambridge University Press 2002) 61, 75; R Murray, *Human Rights in Africa: From the OAU to the African Union* (Cambridge University Press 2004) 51.

[65] Article 5 of the Protocol on the African Court (n 64). See for example, *The Social and Economic Rights Action Centre and the Centre for Economic and Social Rights v Nigeria* (*SERAC*) (n 8); *Tanganyika Law Society and the Legal and Human Rights Centre v The United Republic of Tanzania*, ACtHPR Application No 009/2011, 22 September 2011.

[66] The ACommHPR is empowered to consider cases brought by individuals, groups or NGOs (Article 55 and 56 ACHPR). The ACtHPR has contentious jurisdiction over cases concerning alleged violation of the ACHPR brought by the ACommHPR, States Parties and NGOs (Article 5(1) of the Protocol on the African Court n 64). Individuals are not permitted to bring cases directly to the ACtHPR unless the State Party concerned has made a declaration accepting the jurisdiction of the Court in such matters (Article 5(3) of the Protocol); see J

applicants to be citizens of countries that are party to the Charter.[67] While evidence of violation of the Charter is still necessary, the ACommHPR takes a pragmatic approach and does not require that the actual provisions alleged to have been breached should be specified, as long as there is sufficient information for the Commission or the Court to determine the factual basis of the alleged violation.[68] The ACommHPR and ACtHPR also take a pragmatic approach to the question of evidence, recognising, for example, the importance of media reports in drawing attention to human rights violations. While the ACHPR itself specifies that communications must not be 'based exclusively on news disseminated through the mass media',[69] the Commission has taken the view that reliance may be placed on media reports as one source of information among others, the overall question being one of reliability of the information brought in evidence.[70]

Yet, while the formulation of the rights under the ACHPR and the framework of rules governing claims brought to the ACommHPR and the ACtHPR are much more conducive to environmental claims than are those of either the ECHR or the ACHR, the environmental case law of the African Commission and Court is somewhat sparse. The well-known case of *Social and Economic Rights Action Centre and the Centre for Economic and Social Rights (SERAC) v Nigeria*[71] is to date the only case in which the substantive right to a 'satisfactory' environment has been analysed and applied by the ACommHPR.

The SERAC case concerned complaints of widespread environmental destruction resulting from oil operations in the Niger Delta, including oil spills, contamination of soil and water and other impacts on the homes and health of the people living in the area. The communication alleged that the Government of Nigeria was responsible for violating a number of rights under the ACHPR, including the right to a 'satisfactory' environment,[72] the right to health,[73] and the right to life,[74] through its actions,

---

Harrington, 'The African Court on Human and Peoples' Rights' in M Evans and R Murray (n 64) 305, 318.

[67] In *Maria Baes v Zaire*, Communication 31/89 (1995), a Danish national submitted a communication to the Commission which was declared admissible.

[68] Viljoen (n 64) 70.

[69] Article 56(4) ACHPR.

[70] *Sir Dawda K Jawara v The Gambia*, Communications 147/95 and 149/96 (2000).

[71] *SERAC* (n 8).

[72] Article 24 ACHPR.

[73] Article 16 ACHPR.

[74] Article 4 ACHPR.

the actions of its agents and its failure to take action to conduct environmental impact assessments and to regulate and monitor the activities of the oil companies.[75] The petition was filed by two NGOs on behalf of the Ogoni people who live in the Niger Delta and whose lives, livelihoods and lands were affected.[76]

The case is noteworthy for the extent to which the ACommHPR demonstrates an understanding of the importance of connecting rights, the environment and communities. First, the ACommHPR specifically characterised the communication as an *actio popularis*, noting that this was 'wisely allowed under the African Charter'.[77] This is an important difference between the African system and the other two regional systems, and suggests that there is no inherent reason in the nature of human rights law preventing human rights courts from providing an avenue for consideration of claims of environmental damage brought on behalf of the wider community. Secondly, rather than focusing on isolated rights, the Commission directed attention to the wider impact of oil pollution and the degradation of the environment, linking this wider impact in particular to both the right to a 'satisfactory' environment and the right to health. By linking the right to health and the right to a healthy environment, the Commission was able to give both rights a wide interpretation, concluding that the obligations of the Nigerian State under the ACHPR include the obligation to take action 'to prevent pollution and ecological degradation, to promote conservation and to secure an ecologically sustainable development and use of natural resources'.[78] In the view of the Commission, compliance with both rights also requires that the state itself carries out assessments or permits the independent assessment and monitoring of both the environmental and the social impact of industrial developments.[79]

The third significant aspect of the case was an unequivocal recognition of the intimate connection between rights, even rights that are not

---

75 *SERAC* (n 8) paras 1–9.

76 See KSA Ebeku, 'The Right to a Satisfactory Environment and the African Commission' (2003) 3 *African Human Rights Law Journal* 149; BH Oxman and D Shelton, 'Decision regarding Communication 155/96 (Social and Economic Rights Action Center/Center for Economic and Social Rights v Nigeria)' (2002) 96 *AMJIL* 937.

77 *SERAC* (n 8) para 51.

78 Ibid para 54.

79 Ibid paras 54–55. The Commission also concluded that the Nigerian government was obliged to provide affected communities with information and opportunities to participate in decision-making (para 55, see discussion of procedural environmental rights below).

explicitly protected under the ACHPR.[80] The Commission held that both the right to food and the right to housing or shelter were so closely connected to and interdependent with rights enjoying explicit protection under the Charter that those rights should be read in. The Commission held, for example, that: 'The right to food is inseparably linked to the dignity of human beings and is therefore essential for the enjoyment and fulfilment of such other rights as health, education, work and political participation'.[81]

Finally, the SERAC case demonstrates the approach of the African Commission to group rights.[82] The right to a 'satisfactory' environment is one of a number of group rights protected under the ACHPR. The claim in SERAC related to the rights of the Ogoni people, a minority community whose land and livelihood were impacted by oil extraction in the Niger Delta. The Commission held that the rights of the Ogoni people to a healthy environment[83] as well as their right to the free disposal of wealth and natural resources[84] had been violated. In dealing with both rights, the Commission distinguished between Nigerians as a whole – for whose benefit the Nigerian government is entitled to develop the oil industry – and the rights of the communities who live in the areas affected by oil extraction, and concluded that 'the Ogoni people' as a distinct community are protected by the peoples' rights provided for in the ACHPR.[85]

The subsequent case of *Centre for Minority Rights Development (Kenya) and Minority Rights Group (on behalf of Endorois Welfare Council) v Kenya*[86] endorses the Commission's approach to collective rights and to the interrelationship between rights. The case involved the eviction of the indigenous Endorois community from their ancestral land to make way for a game reserve. The Commission recognised the Endorois as 'a distinct tribal group whose members enjoy and exercise certain rights, such as the right to property, in a distinctly collective

---

80 See Grant (n 12) 166.

81 *SERAC* (n 8) para 68.

82 The ACommHPR specifically notes that 'the importance of community and collective identity in African culture is recognised throughout the African Charter' (*SERAC* (n 8) para 40).

83 Article 24 ACHPR.

84 Article 21 ACHPR.

85 *SERAC* (n 8) para 56.

86 *Centre for Minority Rights Development (Kenya) and Minority Rights Group (on behalf of Endorois Welfare Council) v Kenya (Endorois)*, Communication 276/03.

manner'[87] and found the Kenyan government to be in breach of the Endorois' rights to dispose freely of their natural resources,[88] to culture,[89] to practise religion,[90] to property,[91] and to development.[92]

The *SERAC* and *Endorois* cases demonstrate a clear understanding on the part of the ACommHPR of the importance of the interconnections between environmental degradation and the multiple rights of whole communities, and opens up much greater scope for environmental protection via human rights law under the African system.

Despite the fact that group rights are not clearly endorsed by the ECtHR, there are promising signs of other incremental developments signalling a move away from the limitations of traditional human rights jurisprudence. Although the ECtHR does not permit public interest claims, one area where the jurisprudence of the ECtHR demonstrates acceptance that protection of the environment is a matter of general interest is in relation to the circumstances in which states are permitted to interfere with the right to property. Most rights under the ECHR are not absolute[93] and the right of public authorities to restrict property rights in the 'general interest' is explicitly recognised.[94] In *Hamer v Belgium*,[95] for example, the local authority authorised the demolition of the applicant's house, which had been built in breach of relevant planning legislation. The ECtHR held that state interference with the right to property of the applicant was justified and that there had therefore been no violation of Article 1 Protocol 1. Significantly, in coming to this conclusion, the ECtHR emphasised the importance of protection of the environment in the public interest:

> The environment is a cause whose defence arouses the constant and sustained interest of the public, and consequently the public authorities. Financial imperatives and even certain fundamental rights, such as ownership, should not be afforded priority over environmental protection considerations, in particular when the State has legislated in this regard.[96]

---

87 Ibid para 161.
88 Article 21.
89 Article 7.
90 Article 8.
91 Article 14.
92 Article 22.
93 DJ Harris et al. (n 37) 13.
94 Article 1 of Protocol 1 to the ECHR.
95 *Hamer v Belgium*, Application No 21861/03, 27 November 2007.
96 Ibid para 79.

It is thus clear, that at least in the context of justifying restrictions on property rights, the ECtHR recognises that the collective interest of the public in protection of the environment can trump individual rights.[97]

## 4. COMMUNITIES AND PROCEDURAL ENVIRONMENTAL RIGHTS

The second 'thread' of environmental human rights that has blossomed over the past few decades is often referred to as the 'proceduralisation' of environmental rights.[98] Principle 10 of the Rio Declaration 1992 identified three specific procedural requirements that should apply in situations in which environmental concerns are raised:

> Environmental issues are best handled with participation of all concerned citizens, at the relevant level. At the national level, each individual shall have appropriate access to information concerning the environment that is held by public authorities, including information on hazardous materials and activities in their communities, and the opportunity to participate in decision-making processes. States shall facilitate and encourage public awareness and participation by making information widely available. Effective access to judicial and administrative proceedings, including redress and remedy shall be provided.[99]

The three 'pillars' of procedural environmental rights identified in Principle 10, namely access to information, public participation and access to justice are widely recognised.[100] One of the important features of procedural environmental rights is that they clearly incorporate a community dimension. There is a specific recognition that environmental hazards affect whole communities and it is clearly envisaged that

---

97 Ibid para 80; see also *Depalle v France*, Application No 34044/02, 29 March 2010 and *Brosset-Triboulet and Others v France*, Application No 34078/02, 29 March 2010.

98 M Lee and C Abbot, 'The Usual Suspects? Public Participation under the Aarhus Convention' (2003) 66 *MLR* 80; K Morrow, 'Worth the Paper They Are Written On? Human Rights and the Environment in the Law of England and Wales' (2010) 1 *JHRE* 66, 67.

99 Rio Declaration on Environment and Development, UN Conference on Environment and Development, UN Doc A/CONF.151/26/Rev.1; 31 ILM 874 (1992).

100 M Fitzmaurice, 'Environmental Degradation' in D Moeckli, S Shah and S Sivakumaran, *International Human Rights Law* (Oxford University Press 2010) 622, 636.

provision should be made for all members of such communities to exercise procedural rights. However, implementation across the three regional human rights systems varies.

Under the ECHR, Article 10 guarantees freedom to receive and impart information. In *Steel and Morris v the United* Kingdom,[101] the ECtHR indicated that there is strong public interest in enabling individuals and groups to contribute to the public debate by disseminating information and ideas on matters of general public interest such as the environment. Interference with the right will only be justified if there is a legal basis for the action, in pursuance of a legitimate aim and proportionate to the interest pursued. The Court has taken the view that because participation by interest groups in public debate is essential in a democratic society, the State must show the existence of a 'pressing social need' in order to restrict the right and that the level of protection afforded the right to disseminate information is therefore high.[102]

However, the right to freedom of information under Article 10 is limited to the right to receive and impart information: it does not include a right of access to information or a duty on the part of the state to provide information. In the absence of explicit provision for access to information, the ECtHR has incorporated procedural environmental rights into both Articles 8 and 2. This development, arguably, owes much to the fact that the Convention on Access to Information, Public Participation in Decision-making and Access to Justice in Environmental Matters (Aarhus Convention), which enshrines the principles of access to information, public participation and access to justice, enjoys widespread support in the European context.[103] Indeed, the EU itself is party to the Convention. In *Guerra and Others v Italy*,[104] for example, the applicants complained about toxic emission from a chemical factory situated close to their town. They argued that their rights under both Articles 10 and 8 of the ECHR had been breached as a result of the failure of the authorities to inform the public of the risks and of what steps to take in the event of an accident. The ECtHR rejected the claim under Article 10 on the basis that

---

[101] *Steel and Morris v UK,* Application No 68416/01, 15 February 2005, para 89.

[102] Council of Europe, *Manual on Human Rights and the Environment* (n 32) 79.

[103] Convention on Access to Information, Public Participation in Decision-making and Access to Justice in Environmental Matters, adopted on 25 June 1998, 2161 UNTS 447; 38 *ILM* 517 (1999); see N de Sadaleer, 'Enforcing EUCHR Principles and Fundamental Rights in Environmental Cases' (2012) 81 *Nordic Journal of International Law* 39, 49.

[104] *Guerra and Others v Italy* (n 36).

the right to information did not impose positive obligations on States to provide information,[105] but upheld the complaint under Article 8, holding that a positive obligation to inform the applicants of the risk posed by the factory was imposed as an aspect of the right to respect for private and family life.[106] In *Öneryıldız v Turkey*[107] the ECtHR held that the obligation to provide information established in relation to Article 8 in *Guerra* was similarly applicable to Article 2, the right to life. These cases indicate an important broadening of interpretation of both Articles 8 and 2 to include procedural environmental rights. Moreover, it is significant that in *Öneryıldız* the Court referred to a failure on the part of the administrative authorities 'to inform the inhabitants of the Kazim Karabekir area' of the risks.[108] Thus, although the rights under the ECHR are individualised by the ECtHR, as discussed above, the duty to provide information on environmental risks is not confined to particular individuals – a development clearly recognising that environmental risks affect whole communities and that the right to be informed applies to whole communities.[109] By incorporating procedural rights into Articles 8 and 2, the ECtHR has thus also signalled an understanding of the close connection between procedural and substantive environmental rights.

Similar developments can be seen in the case law of the ECtHR in relation to the right to public participation in the absence of specific provision for this in the ECHR. In *Hatton and Others v UK*,[110] the ECtHR concluded that an enquiry into the decision-making process was required under Article 8 in determining whether the interests of the individual had been given due weight and that public participation was relevant in this context.[111] This approach was applied, for example in *Giacomelli v Italy*[112] where the court held that Article 8 had been breached because of the failure of the authorities to proceed in accordance with procedural safeguards provided for in domestic law, particularly the failure to comply with 'the obligation to conduct an environmental-impact assessment prior to any project with potentially harmful environmental consequences and the possibility for any citizens

---

105 Ibid para 53.

106 Ibid para 60. See also *Tătar v Romania* (n 42).

107 *Öneryıldız v Turkey*, Application No 48939/99, 30 November 2004.

108 Ibid para 87.

109 See also *Hardy and Maile v UK*, Application No 31965/07, 14 February 2012.

110 *Hatton and Others v UK* (n 34).

111 Ibid para 128.

112 *Giacomelli v Italy*, Application No 59909/00, 2 November 2006.

concerned to participate in the licensing procedure and to submit their own observations to the judicial authorities'.[113] It is significant that although the ECtHR has not explicitly required environmental impact assessments to be conducted as an aspect of procedural environmental protection under Article 8, where such assessments are required by domestic law, the court has required that states should implement them and that the public should have access to the results.[114] The ECtHR has similarly extended Article 8 to include the right of access to justice.[115]

Boyle considers the incorporation of environmental procedural requirements into Article 8 to be 'a profound extension of the scope of Article 8' which 'goes far to translate into European human rights law the procedural requirements set out in Principle 10 of the Rio Declaration and elaborated in European environmental treaty law'.[116]

This interpretive extension of rights through the deployment of procedural requirements is not unique, moreover, to the ECtHR. All three pillars of procedural environmental rights have been implemented by the ACommHPR as an integral aspect of both the right to health and the right to a 'satisfactory' environment under Articles 16 and 24 of the ACHPR. In holding that both rights had been violated by the Nigerian state in *SERAC*, the ACommHPR specified a range of obligations inherent to both rights:

> [G]overnment compliance with the spirit of Articles 16 and 24 of the African Charter must also include ordering or at least permitting independent scientific monitoring of threatened environments, requiring and publicising environmental and social impact studies prior to any major industrial development, undertaking appropriate monitoring and providing information to those communities exposed to hazardous materials and activities and providing meaningful opportunities for individuals to be heard and to participate in the development decisions affecting their communities.

In *Endorois* the question of procedural rights was raised in the context of the right to property of indigenous communities and the right to development. The ACommHPR held that even if the creation of a game

---

113 Ibid para 94.

114 Council of Europe, *Manual on Human Rights and the Environment* (n 32) 91.

115 Ibid 101; *Tătar v Romania* (n 42); *Giacomelli v Italy* (n 112); see A Boyle, 'Human Rights or Environmental Rights? A Reassessment' (2007) 18 *Fordham Environmental Law Review* 471.

116 Boyle (n 115) 26.

reserve was a legitimate aim in interfering with the Endorois community's rights to property and development, the State had not made provisions for effective participation of the community in decision-making, no environmental and social impact assessments had been undertaken and there was insufficient provision for compensation or benefit sharing.[117] The ACommHPR thus imposed a three-part procedural 'test', which must be satisfied in order for the interference with the property rights of indigenous communities to be justified.[118] And clearly, the participation required was that of the community, rather than individuals.

The approach of the ACommHPR to the justification of restrictions on the property rights of indigenous peoples, in particular the three-part 'test', was directly derived from the jurisprudence of the IACtHR in *Saramaka People v Suriname*.[119] The Court held there that authorisation by the state of resource exploitation on indigenous land is only justified where the members of the community will share in the benefit; where they have been given an opportunity for effective participation; and only after an independent environmental and social impact assessment has been conducted.[120] The IACtHR has signalled that failure to observe these fundamental environmental safeguards will not only lead to a breach of the right to property of indigenous communities, but also of their right to cultural identity, because of the close bond between culture, community and nature.[121]

As many of the cases discussed in this section demonstrate, developments around procedural environmental rights have largely taken place in the context of resolving the question of whether the State in question had struck an appropriate balance between the interests of the individual and those of the community when placing restrictions on human rights. Since few human rights are defined as being absolute, there is accordingly significant scope for human rights courts to consider ways of accommodating human rights, environmental protection and the interests of

---

[117] *Endorois* (n 86) para 228. See E Ashamu, '*Centre for Minority Rights Development (Kenya) and Minority Rights Group International on Behalf of Endorois Welfare Council v Kenya:* A Landmark Decision from the African Commission' (2011) 55 *Journal of African Law* 300.

[118] The ACommHPR specifically referred to the requirements as a 'test'. *Endorois* (n 86) para 228.

[119] *Saramaka People v Suriname* (n 51).

[120] Ibid para 129.

[121] *Kichwa Indigenous People of Sarayaku v Ecuador* (Merits and Reparations) Inter-American Court of Human Rights Series C No 245 (27 June 2012); see Pavoni (n 11) 33.

communities. However, promising as this development is, and while procedural rights can play an important role in enhancing environmental protection, especially in light of the recognition in all three regional systems that whole communities are entitled to procedural environmental rights, careful assessment of this development is required. The public participation principle, for example, is open to criticism that its implementation does not ensure that the voices of marginalised communities or less powerful groups within communities are heard.[122] It is also arguable that procedural environmental rights are often emphasised at the expense of obliging States to take positive action to address environmental problems. The emphasis on procedural rights, while an important protective strategy, might also be used to deflect attention from a State's failure to act to protect the environment and could operate to place an obligation on individuals to avoid environmental risks, rather than on the State to take action to protect the environment.[123]

## 5. CONNECTING COURTS, TREATIES AND ENVIRONMENTAL HUMAN RIGHTS

One area where interconnection is increasingly evident is between the decisions of human rights courts at international, regional and domestic levels. The terms 'judicial globalisation' and 'transjudicialism', among others, are often used to describe the exchange of ideas between courts, at all levels.[124] National courts have long been one of the primary

---

[122] Lee and Abbot (n 98); S Kuhn, 'Expanding Public Participation is Essential to Environmental Justice and the Democratic Decisionmaking Process' (1999) 25 *Ecology Law Quarterly* 647.

[123] Pavoni (n 11) 85.

[124] AM Slaughter, 'Judicial Globalization' (2000) 40 *Virginia Journal of International Law* 1103; AM Slaughter, 'A Global Community of Courts' (2003) 44 *Harvard International Law Journal* 191; R Bahdi, 'Globalization of Judgment: Transjudicialism and the Five Faces of International Law in Domestic Courts' (2002) 34 *George Washington International Law Review* 555; AM Waters 'Mediating Norms and Identity: The Role of Transnational Judicial Dialogue in Creating and Enforcing International Legal Norms' (2005) 93 *Georgetown Law Journal* 487; A Nollkaemper, *National Courts and the International Rule of Law* (Oxford University Press 2011) 235; A Roberts, 'Comparative International Law? The Role of National Courts in Creating and Enforcing International Law' (2011) 60 *ICLQ* 57.

vehicles for the application of international law in national jurisdictions.[125] However, the global scale and intensity of cross fertilisation of understandings and interpretations between courts and tribunals, domestic, regional and international, has increased enormously over the past few decades, aided – in particular – by the availability of judgments in electronic form.[126]

The interconnections are visible at multiple levels. For example, domestic courts increasingly apply international law, including international environmental principles, relying, amongst other sources, on the jurisprudence of the regional human rights courts. And regional human rights courts take into consideration national constitutional and legislative principles as well as international law principles derived from other international institutions and regional human rights bodies.

One of the formal underpinnings of the application of norms of international law within domestic legal systems derives from the general rules of treaty interpretation of the Vienna Convention on the Law of Treaties (VCLT).[127] National and international courts are required to give effect to international treaties and to interpret them 'in their context and in the light of [their] object and purpose'.[128] Thus, as organs of the state, national courts must prevent their states from acting in contravention of international obligations.[129] A number of other processes support the application of international law norms in domestic courts. Peters notes the increased references to international law and the incorporation of international standards, especially human rights standards, in national constitutions.[130] The South African Constitution[131] illustrates this well. The Bill of Rights in the Constitution is described by Du Plessis as 'an encyclopaedia of international human rights law gleaned from multifarious international declarations, covenants and conventions' as well as

---

125 Nollkaemper (n 124) 7.

126 Slaughter (2000) (n 124) 1109.

127 Vienna Convention on the Law of Treaties (23 May 1969) 1155 UNTS 331.

128 Ibid Articles 29 and 31; see Nollkaemper (n 124).

129 Nollkaemper (n 124) 42; A Peters, 'The Globalization of State Constitutions' in J Nijman and A Nollkaemper (eds), *New Perspectives on the Divide between National and International Law* (Oxford University Press 2007) 251, 300.

130 Peters (n 129) 294–97.

131 Constitution of the Republic of South Africa, 1996.

from the constitutions of other states.[132] The South African Constitution also incorporates the requirement that national law be interpreted in compliance with the state's international obligations.[133] Moreover, section 39(1) requires courts and tribunals, in interpreting the Bill of Rights, to consider international law. The incorporation of international human rights principles into the Constitution, the ongoing constitutional obligation for courts to interpret legislation compliantly with international law, and the requirement to consider international law in interpreting the Bill of Rights, provide important opportunities for the linkages between national law and international human rights law to be recognised in South African jurisprudence. The incorporation of international environmental principles into domestic legislation and the reliance on those principles by the South African Constitutional Court creates further significant connections between national law, international human rights law and international environmental law.[134]

And, as noted above, establishing interconnections between international law and national law is not a one-way process. National constitutional principles as well as domestic human rights and environmental principles have an equally important influence on international law. Peters describes an ongoing circular process of internationalisation of national constitutional principles which are then incorporated back into national constitutions:

> Nowadays, international standards relating to human rights protection, good governance, or even democracy, are frequently incorporated into national constitutions ... Because the origins of those standards frequently lie in domestic constitutional law, the integration of international standards into domestic constitutional law is to some extent the 're-import' of a product which has been modified (sometimes diluted) and which has become more or less universalized in a global discourse.[135]

The case law of the three regional tribunals demonstrates the ongoing role of national law and jurisprudence at the regional level. The modus

---

[132] L du Plessis, 'International Law and the Evolution of (Domestic) Human-Rights Law in Post-1994 South Africa' in Nijman and Nollkaemper (n 129) 309, 312–13.

[133] Constitution of the Republic of South Africa, 1996, section 223.

[134] See *Fuel Retailers Association of Southern Africa v Director-General Environmental Management, Department of Agriculture, Conservation and Environment, Mpumalanga Province and Others*, Case No CCT 67/06, [2007] ZACC 13 (CC); *ILDC* 783 (ZA 2007); see Shelton (n 7) 102.

[135] Peters (n 129) 296.

operandi of the ECtHR when considering cases, for example, is to focus in the first instance on national law, including relevant national constitutional principles. In *Taşkin v Turkey*,[136] for example, the ECtHR considered the claims of violation of Articles 2 and 8 ECHR against the background of national constitutional law and environmental legislation in Turkey. These included the Constitutional right to a 'healthy, balanced environment',[137] the requirement in the Environment Act that an environmental impact report be submitted in relation to projects 'which might cause environmental problems'[138] and the finding of the Supreme Administrative Court, that, based on the environmental impact survey, the operating permit issued to the Ovacik gold mine 'did not serve the public interest'.[139] The Court also referred to relevant international instruments relating to the right to a healthy environment, including Principle 10 of the Rio Declaration[140] and the Aarhus Convention[141] – despite the fact that Turkey was not a party to the Convention, as the Court acknowledged. The ECtHR's finding that Article 8 had been substantively violated was explicitly based on the Supreme Administrative Court's assessment of the balance struck in relation to the applicants' right to a healthy environment and right to life as guaranteed by the Turkish Constitution. This is particularly important for the future development of environmental human rights in light of the increase in the number of national constitutions that now include substantive environmental rights.[142] It suggests that there is considerable scope for national approaches to the implementation of constitutional environmental rights to influence the jurisprudence of regional human rights bodies. The finding that the procedural aspect of the right to private and family life had also been violated was based on the importation of the three pillars of procedural environmental rights into Article 8. Although Principle 10 of the Rio Declaration and the Aarhus Convention were not specifically relied on, the recognition of the relevance of both documents in setting

---

136 *Taşkin v Turkey* (n 43).

137 Article 56 of the Turkish Constitution, see *Taşkin* (n 43) para 90.

138 Section 10 of the Environment Act (Law no 2872); see *Taşkin* (n 43) para 91.

139 *Taşkin* (n 43) para 26.

140 Rio Declaration on Environment and Development (n 99); see *Taşkin* (n 43) para 98.

141 Aarhus Convention (n 103); see *Taşkin* (n 43) para 99.

142 See *Report of the Independent Expert on the issue of human rights obligations relating to the enjoyment of a safe, clean healthy and sustainable environment: Preliminary report* (n 6) para 12.

the context of the case suggests that they were in fact influential in the interpretation of the procedural aspect of Article 8. The approach in *Taşkin*, in which the ECtHR sets out relevant national and international law and jurisprudence before focusing on the claim itself, is typical of the process of investigation followed by the ECtHR and demonstrates the interplay between national constitutional principles and international human rights instruments and jurisprudence.

The connections between the ECHR and other international treaties were specifically discussed in *Demir and Baykara v Turkey*.[143] The ECtHR noted, first, that interpretation is guided by the provisions of the VCLT,[144] and, secondly, that it may also take into account a wide range of relevant rules and principles of international law,[145] including other international treaties, general principles of international law and their interpretation in international and national courts,[146] as well as other Council of Europe instruments and norms.[147] The Court concluded:

> The Court, in defining the meaning of terms and notions in the text of the Convention, can and must take into account elements of international law other than the Convention, the interpretation of such elements by competent organs, and the practice of European states reflecting their common values. The consensus emerging from specialised international instruments and from the practice of Contracting States may constitute a relevant consideration for the Court when it interprets the provisions of the Convention in specific cases. In this context it is not necessary for the respondent State to have ratified the entire collection of instruments that are applicable in respect of the precise subject matter of the case concerned. It will be sufficient for the Court that the relevant international instruments denote a continuous evolution in the norms and principles applied in international law or in the domestic law of the majority of member States of the Council of Europe and show, in a precise area, that there is common ground in modern societies.[148]

---

[143] *Demir and Baykara v Turkey*, Application No 34503/97, 12 November 2008.

[144] Ibid para 65.

[145] Ibid para 67.

[146] Ibid paras 71–73.

[147] Ibid paras 74–75.

[148] Ibid para 85–86; see Harris et al. (n 37) 11–13, noting that the ECtHR now has a research division which provides the Court with assistance regarding comparative and international law; see also Council of Europe, *Manual on Human Rights and the Environment* (n 32) Appendix III: Reference to other instruments relevant to the environment in ECHR case-law, 149.

The ECtHR jurisprudence also demonstrates the extent to which principles of international environmental law have been incorporated into international human rights law. This is particularly evident in relation to the acceptance by the ECtHR of the three pillars of procedural environmental rights, based on the Rio Declaration and the Aarhus Convention and implementation of the precautionary principle in cases such as *Taşkin v Turkey*[149] and *Tătar v Romania*.[150]

The relevance of other international law is specifically dealt with in both the ACHPR and the ACHR. The ACHPR is characteristically expansive in this regard, stipulating that the ACommHPR is to draw inspiration from international human rights law, particularly African human rights instrument, the Universal Declaration of Human Rights (UDHR), other UN human rights treaties and human rights treaties adopted by African countries.[151] The Commission is also directed to 'take into consideration' a range of other sources, including other international treaties, African practice consistent with international human rights norms and 'legal precedents and doctrine'.[152] The ACommHPR has indicated its willingness to implement this, stating in *Purohit and Moore v The Gambia:*

> The African Commission is, therefore, more than willing to accept legal arguments with the support of appropriate and relevant international and regional human rights instruments, principles, norms and standards taking into account the well recognised principle of universality which was established by the Vienna Declaration and Programme of Action of 1993 and which declares that 'all human rights are universal, indivisible, interdependent and interrelated'.[153]

The provisions of the ACHR are not quite as broad, but include protection of rights recognised by domestic law and other treaties of State Parties as well as the American Declaration of the Rights and Duties of Man.[154]

The extent to which the ACommHPR has made connections between international law and regional law is evident from the cases discussed above. For example, *SERAC* demonstrates the importance of both international human rights law and regional comparative jurisprudence in

---

149 *Taşkin v Turkey* (n 43).
150 *Tătar v Romania* (n 42).
151 ACHPR (n 17) Article 60.
152 Ibid Article 61.
153 *Purohit and Moore v The Gambia,* Communication No 241/2001 (2003).
154 ACHR (n 25) Article 29.

interpreting the ACHPR. In dealing with the merits of the case, the ACommHPR began by elaborating the general obligations of states under the ACHPR, as duties to respect, protect, promote and fulfil. These terms are not found in the ACHPR itself, but represent, according to the CommHPR, 'international accepted ideas of the various obligations engendered by human rights'.[155] In interpreting the right to health and the right to a 'satisfactory' environment, the ACommHPR referred specifically to Article 12 of the ICESCR, noting that it requires states to act to improve 'all aspects of environmental and industrial hygiene'.[156] The Commission also invoked decisions of both the IACtHR and the ECtHR in interpreting the right of peoples to freely dispose of their wealth and natural resources under Article 12 of the ACHPR.[157]

The *Endorois* case[158] provides an even more compelling illustration of the use of comparative jurisprudence, especially the jurisprudence of other regional courts, in cases dealt with by the ACommHPR. As noted above, the communication alleged violation of a number of rights under the ACHPR, including the right of the Endorois community to practise their religion,[159] their right to property,[160] to culture,[161] to freely dispose of natural resources[162] and to development.[163] In concluding that all the rights relied on had been violated by the Kenyan State, the ACommHPR referred extensively to decisions of, especially, the IACtHR, but also the ECtHR. The preliminary issue for determination in the case was whether the Endorois should be recognised as a 'distinct community' and if so, whether they were 'indigenous peoples and thereby needing special protection?'[164] In answering these questions in the affirmative, the ACommHPR referred to the case of *Saramaka People v Suriname*[165] as being of 'particular relevance' to the Endorois case[166] and engaged in detailed comparison of the arguments in the two cases before concluding, using the same line of reasoning as the IACtHR, that the Endorois were indeed an indigenous community. The jurisprudence of the IACtHR was

155 *SERAC* (n 8) para 45.
156 Ibid para 54.
157 Ibid para 59.
158 *Endorois* (n 86).
159 ACHPR, Article 8.
160 Ibid Article 14.
161 Ibid Article 17.
162 Ibid Article 21.
163 Ibid Article 22.
164 *Endorois* (n 86) para 146.
165 *Saramaka People v Suriname* (n 51).
166 *Endorois* (n 86) para 159.

similarly influential in determining whether the right to property of the Endorois had been violated, the Commission referring to comparative case law of the IACtHR at each stage of the argument. First, the ACommHPR determined that the 'rights, interests and benefits' of traditional communities in their traditional lands constitute 'property' under the ACHPR and that the rights of the Endorois had been encroached upon, again drawing extensively on *Saramaka*, amongst other cases of the IACtHR,[167] as well as on decisions under the ECHR.[168] The *Saramaka* case was again discussed in detail in determining whether the interference with the Endorois' right to property was in the public interest and in accordance with the law as required by the ACHPR. The jurisprudence of the IACtHR was particularly influential in determining that consultation and compensation were essential elements of the 'in accordance with the law' test. In fact *Saramaka* was the only authority cited in support of the conclusion that effective participation, community benefit and prior environment and social impact assessments were elements of consultation.[169] The ACommHPR also analysed *Saramaka* in detail in dealing with the complaint that the rights of the Endorois to free use of the natural resources had been violated, noting in particular the recognition by the IACtHR of the 'interconnectedness' of thc rights to property, culture and free use of natural resources.[170] The *Endorois* case thus provides ample evidence of the cross-fertilisation of ideas between the regional courts and the indisputable interconnectedness of their jurisprudence.

This brief overview of the legal frameworks within which the regional human rights courts interpret environmental human rights and the practice of the courts, reveals considerable scope for the development of interpretations of human rights that are consistent with protection of the environment.[171] As has been shown, international and comparative law and jurisprudence provide strong support for the incorporation of environmental principles into the regional jurisprudence, even in the absence of provision for substantive environmental rights, as the well-developed procedural environmental rights jurisprudence demonstrates. Moreover, the approach of all three courts is dynamic and contextual, with all three courts recognising the need to interpret legal texts in light of changing circumstances. As Shelton argues:

---

167 Ibid paras 192–97, 205–208.

168 Ibid paras 201–202.

169 Ibid para 227.

170 Ibid paras 257–66.

171 Shelton (n 7) 93.

> Such an approach benefits environmental rights, which might be excluded from consideration using an 'originalist' interpretation of human rights agreements, because most of the latter were written before environmental law developed and thus environmental conditions were not contemplated by the treaty drafters.[172]

## 6. CONCLUSION

Nijman and Nollkaemper argue that human rights norms are increasingly recognised as 'hierarchically higher universal values', evidenced by the fact that the constitutions of a large number of states provide for the supremacy of human rights standards only, a status that applies to no other international standards.[173] Thus, they argue, that although international law itself does not enjoy supremacy in domestic legal systems, universal values, such as human rights standards are often recognised as superior. This hierarchy arises because of the recognition of the moral force of human rights, within both the international and national legal systems. The new hierarchy in development is thus not a formal one in which international law trumps national law, but one of substance, in which certain values are viewed as being above all others.[174] But as Nijman and Nollkaemper point out, the recognition of superior values or norms is challenged by a multiplicity of competing perspectives and is therefore merely one understanding of current dynamics within international law.[175]

The role of litigation should also not be overstated. Courts are only one of a number of important players able to shape national and international approaches to environmental protection. But it should also be recognised that the influence of litigation potentially extends far beyond the outcome of any particular case. Gearty draws attention to the moral weight carried by human rights and the ways in which the language of human rights can be used in order to influence the wider debate about environmental protection: '[Y]ou do not have to succeed in achieving a new law to have triumphed in your deployment of the language of human rights. The legal arena is not the only test of utility'.[176]

---

172 Ibid 94.

173 J Nijman and A Nollkaemper, 'Beyond the Divide' in Nijman and Nollkaemper (n 129) 342.

174 Ibid.

175 Ibid 347.

176 Gearty (n 10) 20.

Many of the participants at the Oñati Workshop argued for bold innovation – for a sweeping aside of existing paradigms. Realism however dictates incremental steps.[177] This analysis suggests that the seeds of a new paradigm are being sown and new ways of doing things are being tested, especially in approaches seen in the African and Inter-American system. Courts are well versed in policy development, but they operate in a particular political context and are constrained by institutional and functional factors.[178] The difficulties should not be underestimated, including the perceived need to show deference to national authorities and lack of political will to implement certain measures, to mention but two of the most fundamental problems. However, the analysis presented here demonstrates how values already inherent in the system are available to be used to take forward the development of environmental protection within human rights law. We have the tools. But the arguments and techniques identified in the jurisprudence of the regional human rights courts must be disseminated more widely and become mainstream. From this perspective, criticisms of the individualism and disconnection of human rights law may be – for all its necessity – overstated, and re-imagining environmental human rights adjudication may not be that radical after all.

---

177 See HM Osofsky, 'Climate Change and Environmental Justice: Reflections on Litigation Over Oil Extraction and Rights Violations in Nigeria' (2010) 2 *JHRE* 189, 209.

178 Shelton (n 7) 118.

# PART 3

# Activism and praxis

# 7. Human rights practice: a means to environmental ends?

***Kate Donald****

## 1. INTRODUCTION

The symbiotic relationship between enjoyment of human rights and environmental quality is now well established.[1] Although until recently the human rights and environmental movements operated largely in isolation from each other, relationships of cooperation and collaboration are growing. There is now a strong current within international human rights organisations, institutions and mechanisms to incorporate environmental issues within their work.[2] In terms of the foci of activism and advocacy,

* This chapter is adapted from an article first published as K Donald, 'Human Rights Practice: A Means to Environmental Ends?' (2013) 5/3 *Oñati Socio-Legal Series* 908–30.

Kate Donald is a human rights researcher and advocate, currently Director of the Human Rights in Development programme at the Center for Economic and Social Rights (cesr.org) in New York. (This article does not necessarily reflect the views of CESR.) From 2012 to 2014 she was Adviser to the United Nations Special Rapporteur on extreme poverty and human rights; previously she held positions at the International Council on Human Rights Policy, and at the UN Office of the High Commissioner for Human Rights in Geneva. Kate has an interdisciplinary Master's degree in Human Rights from the London School of Economics, and a Bachelor's degree in Modern History from the University of Oxford.

[1] See for example, International Council on Human Rights Policy, *Human Rights and Climate Change: A Rough Guide* (ICHRP 2008) and Council of Europe, *Manual on Human Rights and the Environment* (2nd edn, Council of Europe Publishing 2012); JH Knox, 'Report of the Independent Expert on the issue of human rights obligations relating to the enjoyment of a safe, clean, healthy and sustainable environment: Mapping report' (2013) UN Doc A/HRC/25/53.

[2] See for example, Human Rights Watch, http://www.hrw.org/topic/environment (accessed 16 November 2014), and Office of the High Commissioner

this trend can be divided into three different types of approach, each premised on a somewhat different understanding of the nature and strengths of human rights practice:

- Lobbying for the adoption and institutionalisation of new environmental human rights;
- Using existing human rights mechanisms to tackle environmental harms;
- Taking a human rights-based approach to environmental practice.

This chapter will focus on how these three approaches are being realised, developed or used in practice, before going on to discuss potential alternatives and examples of radical or hybrid approaches. A 'principled strategy' for the embedding of the interdependence of human rights and the environment in human rights practice will be discussed, based on a realistic interpretation of the strengths and weaknesses of human rights in its various shifting and contested forms. Finally, the chapter will survey the challenges present at the level of international institutions and policy-making outside of the human rights sphere.

As part of the analysis, comparisons will be made with development, as a field of theory and practice that has been the subject of human rights approaches and standard-setting for some time. Activism and praxis at the international level will be the main focus, in particular that using or targeting UN human rights mechanisms, but national and grassroots practice will also be included in the analysis. Legal approaches are included, as one of the current dominant forms of human rights practice.

## 2. ENVIRONMENTAL HUMAN RIGHTS

Many human rights practitioners and environmental activists are now involved in advocating for the institutionalisation of substantive environmental human rights (EHR), and/or use the language of EHR.[3] This

---

for Human Rights, http://www.ohchr.org/EN/Issues/HRAndClimateChange/Pages/HRClimateChangeIndex.aspx (accessed 16 November 2014). In terms of strategic litigation, see for example, *Social and Economic Rights Action Center and Center for Economic and Social Rights v Nigeria*, African Commission on Human and People's Rights, Communication No 155/96.

[3] In terms of organisations, see for example the Center for International Environmental Law, Friends of the Earth, The Center for Human Rights and the Environment, Oxfam, Earthjustice. In 2003 Jan Hancock conducted a survey of

section analyses arguments that EHR should be enshrined in a new human rights treaty or other legal standard, not the more fluid, less legalistic claims linking human rights and the environment emanating from social movements (which will be discussed in section 4).

The concept of EHR has been emerging for some time, at least since the 1972 Stockholm Declaration,[4] which was the first authoritative statement to formulate international obligations to protect the environment in the language of human rights. It declared that: 'Man has the fundamental right to freedom, equality, and adequate conditions of life, in an environment of a quality that permits a life of dignity and well-being, and he bears a solemn responsibility to protect and improve the environment for present and future generations'.

Subsequent international standard-setting documents and international/regional human rights instruments have included rights to 'an environment adequate for health and well-being',[5] to 'a general satisfactory environment favourable to [human] development',[6] and to 'live in a healthy environment'.[7] In 1994, the UN's 'Ksentini Report' included a Draft Declaration of Principles on Human Rights and the Environment. It declared that 'human rights, an ecologically sound environment, sustainable development and peace are interdependent and indivisible' (Principle 1) and postulated 'the right to a secure, healthy, and ecologically sound environment' (Principle 2) and 'the right to an environment adequate to meet equitably the needs of present generations and that does not impair the rights of future generations to meet equitably their needs'

---

NGOs (chosen by him as representative of the 'environmental social movement'), which were asked whether they recognised EHR. Of these, 54 answered affirmatively, 36 declined to state a position and only 3 stated that they did not. J Hancock, *Environmental Human Rights: Power, Ethics, and Law* (Ashgate 2003) 56.

4 Declaration of the United Nations Conference on the Human Environment (adopted 16 June 1972) UN Doc.A/CONF.48/14, at 2 and Corr.1.

5 World Commission on Environment and Development (UNWCED), *Our Common Future* (*Brundtland Report*) (Oxford University Press, 1987) 348.

6 African Charter on Human and People's Rights (adopted 27 June 1981, entered into force 21 October 1986) OAU Doc. CAB/LEG/67/3 rev. 5, 21 I.L.M. 581981, art 24.

7 Additional Protocol to the American Convention on Human Rights in the Area of Economic, Social and Cultural Rights 'Protocol of San Salvador' (adopted 17 November 1988, entered into force 16 November 1999) OAS A-52, art 11.

(Principle 4).[8] The report also concluded that there had been a 'shift from environmental law to the right to a healthy and decent environment'[9] and that this right was part of international law and capable of immediate implementation by existing human rights bodies.[10] In 2011, by May and Daly's estimation, nearly sixty states had environmental rights embedded in their constitution.[11]

EHR are articulated in many different forms, as illustrated by the above examples and the varying definitions many academics have given to EHR. Henry Shue classifies access to the atmosphere's 'emission absorptive capacity' as being a basic right and suggests that every person ought to have the right to a basic level of GHG emissions.[12] Simon Caney argues that persons have a human right to a healthy environment, a 'human right not to suffer from the ill-effects of global climate change'[13] and that 'those who contribute to global climate change through high emissions are guilty of human rights violations'.[14] Steve Vanderheiden posits the 'right to climatic stability' and also endorses 'survival emissions' as a basic right.[15]

A rigorous analysis of the ontological or metaphysical foundations of these various conceptions of EHR is outside the scope of this chapter;[16] rather, this chapter concentrates on whether advocating for EHR is strategically useful or desirable. Firstly, many environmental activists argue that such a strategy is *not* desirable, due to the different goals of the environmental and human rights movements. One may well ask the 'critical question' of 'whether the environment is really very well served

---

[8] Fatma Ksentini, *Final Report of the UN Sub-Commission on Human Rights and the Environment, UN Doc.E/CN/Sub.2/1994/9* (United Nations, 1994) Annex 1.

[9] Ibid para 22.

[10] Ibid paras 242 and 258.

[11] JR May and E Daly, 'New Directions in Earth Rights, Environmental Rights and Human Rights: Six Facets of Constitutionally Embedded Environmental Rights Worldwide' [2011] *IUCN Academy of Environmental Law eJournal.*

[12] S Vanderheiden, *Atmospheric Justice: A Political Theory of Climate Change* (Oxford University Press 2008) 247.

[13] S Caney, 'Human Rights, Climate Change, and Discounting' (2008) 17(4) *Environmental Politics* 536–55.

[14] S Caney, 'Cosmopolitan Justice, Rights and Global Climate Change' (2006) 19(2) *Canadian Journal of Law and Jurisprudence* 255, 278.

[15] Vanderheiden (n 12) 241–43.

[16] See the work of Henry Shue, Stephen Gardiner, Simon Caney, Richard Hiskes and Jan Hancock.

by enhancing the right of *humans*'.[17] Indeed, Arne Naess professes dismay at the 'mass of ecologically irresponsible proclamations of human rights' and environmental ethicist John Rodman sees rights as being tied up with the 'project of modernity – the total conquest of nature'.[18] James Lovelock, at the extreme, sees the human species locked in rapacious conflict with the Earth and predicts a time when Earth's 'plague of people' is 'cured' by the death of billions.[19] Equally, human rights activists criticise the environmental movement for disregarding immediate human needs. Anderson argues that such tensions cannot be wished away, and 'the fashionable view' that human rights and environmental protection are interdependent and compatible 'serves as a moral comforter which temporarily cloaks the extremely difficult questions which must be faced'.[20] It is indeed difficult to see how a human rights treaty with universal applicability could satisfactorily overcome these concerns and weave these values together.

However, the strategic appeal of advocating for environmental human rights is also clear. As Woods states, 'the idea of EHR offers an opportunity to critically engage with what is undoubtedly the biggest game in town in terms of moral language used in politics'.[21] Summarising his conclusions after a survey of NGOs working on environmental issues, Hancock observes that EHR claims are seen as 'a useful tactical approach for achieving political change', largely because linking human rights and environmental claims is perceived to broaden the social base and appeal of a campaign.[22] The main perceived advantage of institutionalising EHR is the strength of the claim, the 'trump' effect that allegedly puts human rights claims beyond political compromise. There is a clear desire to increase opportunities for legal protection of the environment by

---

17 T Hayward, *Constitutional Environmental Rights* (Oxford University Press 2005) 32.

18 Quoted in W Aiken, 'Human Rights in an Ecological Era' (1992) 1 *Environmental Values*, 191.

19 Quoted in J Gray, *Straw Dogs: Thoughts on Humans and Other Animals* (Granta 2002) 6.

20 M Anderson, 'Human Rights Approaches to Environmental Protection: An Overview' in AE Boyle and MR Anderson (eds), *Human Rights Approaches to Environmental Protection* (Oxford University Press 1996) 3.

21 K Woods, 'What Does the Language of Human Rights Bring to Campaigns for Environmental Justice?' (2006) 15(4) *Environmental Politics* 572, 579.

22 Hancock (n 3) 62.

harnessing the power of human rights law and its enforcement mechanisms.[23] As Nickel says, 'the human rights movement has strong international recognition, support, and institutions and thus has valuable resources to offer environmentalism'.[24]

However, despite the understandable appeal of human rights law, advocating for an international treaty enshrining EHR may well be misguided, as institutionalised EHR would manifest a number of notable weaknesses. Given that climate change and environmental degradation manifest themselves across borders and that business entities play a large role in creating and exacerbating these harms, issues of environmental justice require a rigorous engagement with power balances within and between states, and with corporate power. Many critics claim that institutionalised human rights are frequently used to defend the status quo and to resist radical action. Stammers, for example, argues that although initially rights claims can challenge existing power structures, '[o]nce rights are institutionalized, then they play a highly ambivalent role in respect of power'.[25] Moreover, the legal human rights regime is firmly premised on states as the guarantors and potential violators of human rights and so far has struggled to stretch beyond this somewhat outdated paradigm. While activists have sought to hold governments accountable for their support of corporate polluters,[26] efforts to ensure legal accountability of corporations for human rights violations have been largely fruitless.[27] Indeed, human rights in their legal form can play a conservative role in respect to corporate power, providing an illusory promise of justice while in fact leaving corporate impunity and the pursuit of profit at the expense of human lives and livelihoods unchallenged; even in some cases casting corporations as persons with (human)

---

[23] Woods (n 21) 577.

[24] J Nickel, 'The Human Right to a Safe Environment: Philosophical Perspectives on its Scope and Justification' (1993) 18 *Yale Journal of International Law* 281, 283.

[25] N Stammers, 'Social Movements and the Social Construction of Human Rights' (1999) 21 *Human Rights Quarterly* 980, 999.

[26] See for example, P Newell, 'Climate Change, Human Rights and Corporate Accountability' in S Humphreys (ed.), *Human Rights and Climate Change* (Cambridge University Press 2010).

[27] Although a potentially important breakthrough came in June 2014 when the UN Human Rights Council adopted a resolution (26/9) initiating a process to develop legally binding human rights standards for transnational corporations and other business enterprises.

rights to be protected.[28] Advocates of EHR should thus be wary of envisaging human rights as a reliable weapon to wield against the political, corporate and financial power lined up against tough environmental protection measures.

It is also necessary to acknowledge that the extra-territorial applicability of human rights obligations is highly contested and resisted by states, despite the best efforts of advocates and legal experts.[29] This could clearly be a huge deficiency in terms of EHR claims. Environmental (and especially climate change) harms are often geographically (not to mention temporally) dispersed and the violator is often not the host state of the victims. Certainly, it is not emissions from their own countries that primarily jeopardise the citizens of states that are highly vulnerable to climate-change-related catastrophes like Bangladesh or The Maldives. Additionally, EHR are unlikely to address, and indeed may draw attention away from, global economic problems or other structural causes of climate change and environmental damage. Rights remedies in their legal form often (although not always or inevitably) tackle symptoms rather than causes, which in this case would make it extremely unlikely that courts and human rights mechanisms would consider vital issues such as relationships of political economy, technology choice, forms of production and distribution of goods.[30]

Additionally, there is the question of the technical competence of a human rights mechanism for dealing with environmental issues. As Boyle points out, international law is already replete with rules, principles and criteria for ensuring environmental quality, and it is therefore 'far from certain whether much would be added by reformulating these rules in explicit human rights terms'.[31] All modern environmental treaties have their own supervisory institutions, with more technical competence for weighing environmental claims. Indeed, Handl argues persuasively that human rights mechanisms would be an inefficient, piecemeal route to

---

[28] U Baxi, 'Writing about Impunity and Environment: The 'Silver Jubilee' of the Bhopal Catastrophe' (2010) 1/1 *Journal of Human Rights and the Environment* 23; A Grear, 'Challenging Corporate Humanity: Legal Disembodiment, Embodiment and Human Rights' (2008) 7/3 *Human Rights Law Review*, 511.

[29] See for example, the Maastricht Principles on Extraterritorial Obligations of States in the area of Economic, Social and Cultural Rights. Available at http://www.etoconsortium.org/en/library/maastricht-principles/ (accessed 16 November 2014).

[30] Anderson (n 20) 22.

[31] A Boyle, 'The Role of International Human Rights Law in the Protection of the Environment' in Boyle and Anderson (n 20) 56.

environmental protection, because general standards would have to be developed on a case-by-case basis in reaction to individual complaints, while the environmental interests to be protected are better conceptualised as collective.[32] Additionally, in practice the human rights regime has proven more effective in mandating remedies once a harm has occurred than in ensuring prevention of harm. In the case of climate change, an after-the-fact remedy is clearly insufficient, particularly because the harm is likely to be irreversible and to have already impacted an extraordinarily large number of people.

In order for any legal rights-claim to be judged, a violation of one or more specific rights must be pinpointed. If conceived in broad, holistic terms, EHR would likely be too vague and complex for adjudication or enforcement by human rights bodies, but a list of specific, narrow environmental human rights are unlikely to adequately encapsulate the multi-faceted, context-dependent and inter-linked nature of the impact of environmental degradation on human life. Thus, while claims-making with regard to environmental rights may be a powerful tool for activists (see section 4), it is the contention here that an institutionalised, legal form of environmental human rights does not have great potential to protect and promote both environmental *and* human flourishing – at least not on the time-scale necessary.

## 3. USING EXISTING HUMAN RIGHTS MECHANISMS TO TACKLE ENVIRONMENTAL HARMS

In recent years, UN human rights mechanisms have begun to take environmental questions seriously. In March 2008, the UN Human Rights Council (HRC), 'concerned' by the 'threat' of climate change and its 'implications for the full enjoyment of human rights', issued Resolution 7/23 requesting the Office of the High Commissioner for Human Rights (OHCHR) to conduct a detailed analytical study of the relationship between human rights and climate change.[33] In 2009, HRC Resolution

[32] G Handl, 'Human Rights and the Protection of the Environment: A Mildly "Revisionist" View' in A Trindade (ed.), *Human Rights, Sustainable Development and the Environment* (Instituto Interamericano de Derechos Humanos 1992) 117.

[33] UN Human Rights Council, *Report of the Office of the United Nations High Commissioner for Human Rights on the relationship between climate change and human rights A/HRC/10/61* (United Nations 2009).

10/4 was adopted, '[a]ffirming that human rights obligations and commitments have the potential to inform and strengthen international and national policy-making in the area of climate change'. The topic of the 2010 Social Forum[34] was climate change, and in early 2012 OHCHR held a seminar on human rights and climate change, also at the HRC's direction.[35] For some time, OHCHR had a dedicated member of staff who acted as a focal point on the environment and climate change, but this post has apparently been eliminated or diluted due to budget cuts.[36]

### 3.1 UN Special Procedures

In 2012, after significant civil society lobbying, the HRC announced the appointment of a new UN special procedure, in the form of the 'Independent Expert on the issue of human rights obligation related to the enjoyment of a safe, clean, healthy and sustainable environment' (hereafter, 'I.E. environment'). An American law professor, John Knox, was subsequently appointed as the first mandate-holder. Remarkably, climate change was not explicitly included in the mandate[37] and there are now civil society actors and states advocating for another special procedure on climate change.[38] The introduction of this special procedure illustrates well the way in which existing human rights mechanisms have dealt with the environment 'issue', and is discussed below.

The UN special procedures are a human rights mechanism, created in 1979, that sits alongside the treaty bodies. They are created by the Human Rights Council and staffed by independent experts,[39] or 'working

---

[34] A subsidiary body of the Human Rights Council, serving as a space for interactive dialogue between states, civil society and intergovernmental organizations.

[35] See http://www.ohchr.org/EN/Issues/HRAndClimateChange/Pages/HRClimateChangeIndex.aspx – (accessed 16 November 2014).

[36] Press conference by the United Nations High Commissioner for Human Rights, Zeid Ra'ad Al Hussein, 16 October 2014. Available at http://www.ohchr.org/EN/NewsEvents/Pages/DisplayNews.aspx?NewsID=15173&LangID=E#sthash.c9v117lj.dpuf (accessed 16 November 2014).

[37] See the Human Rights Council resolution creating the mandate A/HRC/19/L.8/Rev.1. Available at http://www.ciel.org/Publications/UNRes_HRandENV_March2012.pdf (accessed 16 November 2014).

[38] There is already (since 1995), a special procedure on the 'human rights obligations related to environmentally sound management and disposal of hazardous substances and waste'.

[39] Variously called 'Independent Experts', 'Special Rapporteurs' or 'Special Representatives of the Secretary General'.

groups' of independent experts, who serve in a voluntary unpaid capacity and are appointed to focus on certain thematic issues or country situations. They are supported by OHCHR, but capacity has grown ever more limited as the number of special procedures has multiplied without corresponding financial support from states.[40] This unwillingness of states to support mandates financially, even as they propose and approve their creation, raises the question of whether the appointment of a special procedure is just an easy conscience-salving 'action' from states on an issue of concern. (One could proffer by way of example, the appointment of a Special Rapporteur on 'the situation of human rights in the Syrian Arab Republic' in 2011.)

The thematic special procedures include mandates on the right to health, on cultural rights, on freedom of expression, on violence against women, on torture, and on extreme poverty. The special procedures undertake country missions, providing recommendations on the implementation of national laws, policies and programmes related to their mandate, and also submit thematic reports on global issues to the UN General Assembly and Human Rights Council.[41] As such, they operate at a high policy level, can exert a degree of influence on states and their reports can be a useful awareness-raising and advocacy tool.[42] They often work closely with civil society and thus are able to raise emerging, controversial or neglected issues in high-level UN forums. However, special procedures are also somewhat constrained in their actions: country visits depend on the invitation of states and diplomacy is generally favoured over naming and shaming.[43] Their recommendations are not legally binding, but may be referred to in Universal Periodic Review[44] sessions and concluding recommendations of treaty bodies, although coordination with these other mechanisms requires systematic

---

[40] T Piccone, *Catalysts for Rights: The Unique Contribution of the U.N.'s Independent Experts on Human Rights* (Brookings Institution 2010); Office of the UN High Commissioner for Human Rights (OHCHR), *United Nations Human Rights Appeal 2013* (United Nations 2013).

[41] See OHCHR website for more information: http://www.ohchr.org/en/HRBodies/SP/Pages/Welcomepage.aspx (accessed 16 November 2014).

[42] See Piccone (n 40) and J Donnelly, *Universal Human Rights in Theory and Practice* (3rd edn, Cornell University Press 2013) 163–4.

[43] P Pinheiro, 'Musings of a UN Special Rapporteur on Human Rights' (2003) 9(1) *Global Governance* 7.

[44] The UPR is a mechanism by which every UN member state submits to peer review (i.e., by other states) of its human rights record every four years (since 2008).

strengthening.[45] Of course, much also depends on the personal commitment and expertise of the individual mandate holder.

The new special procedure on the environment may well be able to have a positive impact. Of course, many academics and activists have already clarified how existing human rights laws, norms and principles apply to environmental harms (as indeed have UN agencies and bodies such as OHCHR and the Committee on Economic, Social and Cultural Rights (CESCR)), but the work of a special procedure can increase visibility and catalyse improvements in respect for human rights, as well as contribute towards standard-setting.[46] The I.E. environment mandate is framed very conservatively ('on *the issue of* human rights obligation *related to* the enjoyment of a safe, clean, healthy and sustainable environment'), but many special procedure mandate holders have been able to push their mandate beyond the terms originally envisaged, albeit with resistance and sometimes backlash from some states.[47] Many advocates hope that the mandate will lay the foundation for the HRC's recognition of the right to a healthy environment.[48]

However, there are also a number of potential weaknesses and drawbacks, on top of the general limitations of special procedures described abovc. It is possible that a special procedure may serve as a panacea, allowing the members of the Human Rights Council to believe that they have taken adequate action on the environment issue. Ultimately, direct and meaningful action on the environment and climate change is unlikely to originate from the Human Rights Council on the urgent timescale required. After all, an environmental mandate has only just been created, and a mandate for climate change remains lacking, despite the links between human rights and the environment having been recognised by

---

[45] M Sepúlveda, 'Report of the Special Rapporteur on extreme poverty and human rights' (2012) UN Doc A/HRC/20/25.

[46] See Piccone (n 40) for analysis and examples.

[47] See for example, E Lee and DJ Horndrup, 'The Right to Health: An Interview with Professor Paul Hunt' (2005) 2/1 *Essex Human Rights Review* 57; and Human Rights Law Centre, 'UN Special Rapporteur Condemns Cruel, Inhuman and Degrading Treatment and Torture in Healthcare' (Human Rights Law Centre website), available at http://www.hrlc.org.au/un-special-rapporteur-condemns-cruel-inhuman-and-degrading-treatment-and-torture-in-healthcare (accessed 29 October 2013).

[48] Center for International Environmental Law (CIEL), 'UN Human Rights Council Establishes Independent Expert on Human Rights and the Environment' (CIEL website 2012), available at http://www.ciel.org/HR_Envir/UN_Resolution_23Mar2012.html (accessed 16 November 2014).

UN bodies since the 1970s (see above) and the Human Rights Commission (the Council's predecessor) having officially recognised the threat of climate change in 1994.[49]

There is also the question of rights fragmentation; there is now a special procedure on toxic waste, on human rights obligations related to the environment, and next, quite possibly, climate change. It is difficult to see how these mechanisms will refrain from overlapping significantly. There is perhaps no harm in this in a strategic sense, but there are wider questions to be asked about the propensity of the human rights regime to divide issues into silos, to damaging effect (for example, the persistent distinction made between economic, social and cultural rights and civil and political rights).

## 3.2 Treaty Bodies and Courts

Clearly, however, special procedures are not the only human rights mechanism that can be applied to tackle environmental issues. Courts and treaty bodies may still play a role, and in some cases already have (see below).

One potentially important new tool is the Optional Protocol to the International Covenant on Economic, Social and Cultural Rights (OP-ICESCR), which came into force in May 2013. The OP-ICESCR allows individuals or groups of individuals to submit complaints to the Committee, a capacity that it had lacked until this point, unlike many other mechanisms (showing the priority traditionally accorded to civil and political rights). The Committee will also be able to launch an inquiry procedure against a State Party (albeit only one that has recognised its competency to do so) if it receives 'reliable information indicating grave or systematic violations by a State Party of any of the economic, social and cultural rights set forth in the Covenant'.[50] This could open up a space for environmental issues to be the subject of Committee investigations and (non-binding) findings, through complaints related to the rights to health, housing, work, adequate standard of living, etc. This will constitute an important and interesting development in terms of supporting accountability for environmental harms, although of course general concerns regarding the potential restrictions of the human rights law approach still stand – regarding, for example, extraterritorial obligations and non-state actors as discussed above.

---

[49] Ksentini (n 8).

[50] UN General Assembly, *Resolution 63/117* (United Nations 2008).

Of course, environmental issues do not only engage social and economic rights, as the UN Economic Commission for Europe's Convention on Access to Information, Public Participation in Decision-making and Access to Justice in Environmental Matters (Aarhus Convention)[51] explicitly recognises. Indeed, most cases regarding environmental quality or degradation have been given court hearings as civil and political rights claims.[52] Courts have tackled issues related to the environment through claims brought on the basis of, for example, the rights to private and family life,[53] effective remedy,[54] life,[55] freedom of expression,[56] property,[57] and health.[58] The Indian Supreme Court has notably used constitutional rights to address issues of environmental protection,[59] while the European Court of Human Rights (ECtHR), despite the lack of environmental, economic or social rights in the Convention, has increasingly examined complaints in which individuals have argued that a breach of one of their Convention rights has resulted from environmental factors.[60] The African Commission on Human and Peoples' Rights has found

---

[51] Convention on Access to Information, Public Participation in Decision-making and Access to Justice in Environmental Matters (adopted 25 June 1998, entered into force 30 October 2001) 2161UNTS 447 (Aarhus Convention).

[52] It is telling that although environmental issues are often classified as within the social and economic rights ambit, the vast majority of existing case law related to the environment is concerned with civil and political rights claims. This partly reflects the fact that social and economic rights are still (falsely) perceived by some to be non-justiciable and aspirational (on justiciability, see A Nolan, B Porter and M Langford, 'The Justiciability of Social and Economic Rights: An Updated Appraisal' (2007) Center for Human Rights and Global Justice Working Paper Number 15). Those bringing environmental claims have been forced to do so within the framework of civil and political rights, or may have strategically chosen to use civil and political rights claims in other cases, recognising their perceived stronger force.

[53] *Lopez Ostra v Spain* 16798/90 [1994] ECHR 46 (9 December 1994).

[54] *Zander v Sweden* [1993] IIHRL 103 (25 November 1993).

[55] *Oneryildiz v Turkey*, 48939/99 [2004] ECHR 657 (30 November 2004).

[56] *Bladet Tromso and Stensaas v Norway*, 21980/93 [1999] ECHR 29 (20 May 1999).

[57] *Maya Indigenous Community of the Toledo District v Belize*, Case 12.053, Report No. 40/04, IACHR, OEA/Ser.L/V/II.122 Doc. 5 rev. 1 at 727 (2004).

[58] *Menores Comunidad Paynemils/acción de amparo*, Cámara de Apelaciones en lo Civil de Neuquén, Sala II, Argentina (19 May 1997).

[59] G Sahu, 'Implications of Indian Supreme Court's Innovations for Environmental Jurisprudence' (2008) 4/1 *Law Environment and Development Journal* 1.

[60] Council of Europe (n 1).

violations under Article 24[61] of the African Convention, for example in the landmark *Ogoniland* case, holding that Article 24 imposes an obligation on the state to take reasonable measures 'to prevent pollution and ecological degradation, to promote conservation, and to secure ecologically sustainable development and use of natural resources'.[62] Such jurisprudence will likely continue[63] but ultimately can only tackle individual (or in some jurisdictions, group)[64] harms caused by environmental damage in a relatively piecemeal way.[65]

Ultimately, tackling environmental issues through human rights mechanisms (be they courts, UN treaty bodies or special procedures) is a worthwhile avenue in certain cases, but one with limited capacity and reach.[66] The new UN special procedure should help in terms of standard-setting, create publicity for the environment/human rights interface and for specific cases of rights violations related to the environment, and may produce useful policy recommendations for states. However, for the special procedure and other human rights mechanisms, enforcement remains a critical gap that makes unlikely any radical or transformative effect on the ground (especially given the powerful vested interests involved), while the narrow focus of treaty bodies and courts on individual cases or national contexts limits the scope of their findings or pronouncements.

---

[61] The right to 'a general satisfactory environment favourable to their development'.

[62] *The Social and Economic Rights Action Center v Nigeria*, African Commission on Human and Peoples' Rights, 155/96 (2001).

[63] A Boyle, *Human Rights and the Environment: A Reassessment. Paper given at UNEP High Level Expert Meeting on the New Future of Human Rights and Environment* (2010).

[64] For instance, the African human rights regime explicitly protects the rights of 'peoples' as well as individuals. See African Charter on Human and People's Rights (n 6).

[65] It should be noted that one necessary area of fusion between human rights and environmental issues is the frequent rights violations that environmental activists experience in many parts of the world, including arbitrary arrest, torture and assassination. Such cases easily fit under the mainstream and regular activities of human rights courts and litigation, being concerned with violations of individuals' rights. Of interest here is the potential for human rights treaty bodies and courts to tackle environmental degradation or climate change more broadly.

[66] Handl (n 32); Boyle (n 31).

## 4. THE HUMAN RIGHTS-BASED APPROACH

In this section, it is argued that the 'human rights-based approach' to environmental matters has greater potential to be far-reaching and holistic than the products of formal human rights mechanisms. When applied to activism and action on the ground, with a sensitivity to context, these approaches can engage more effectively with the structural and multi-faceted causes of rights violations and lack of rights enjoyment.

### 4.1 Components of a Human Rights-based Approach

There is no single definition of the human rights-based approach (alternatively called rights-based approach, or RBA). However the core elements, common across nearly all frameworks, include: accountability; equality and non-discrimination; participation and empowerment.[67] None of the elements are in themselves revolutionary concepts but if implemented together, meaningfully, the potential for more sustained and sustainable change is increased. It is important to note that the understanding of each of the components of the RBA is quite particular, based on human rights obligations enshrined in international treaties and courts' and treaty bodies' interpretation of said rights. For example, under the RBA the principles of equality and non-discrimination would mandate that all measures (say, environmental protection or climate change mitigation measures) should be implemented in a gender-sensitive, non-discriminatory manner that ensures that all concerned individuals and groups can benefit equally. More progressively, substantive equality under human rights law also implies affirmative action, so disadvantaged individuals and groups should be identified and *prioritised* in any measure taken, including in resources allocated.[68] Given that climate change and environmental degradation have a disproportionate effect on

---

[67] See UN Development Group, *The Human Rights Based Approach to Development Cooperation: Towards a Common Understanding among UN Agencies* (United Nations 2003).

[68] See for example, Convention on the Elimination of All Forms of Discrimination against Women (adopted 18 December 1979, entered into force 3 September 1981) 1249 UNTS 13, art 4(1); and International Convention on the Elimination of All Forms of Racial Discrimination (adopted 7 March 1966, entered into force 4 January 1969) 660 UNTS 195, art 2(2). Treaty bodies have consistently interpreted non-discrimination and equality along these lines: see for example, UNCHR 'General Comment 18' in 'Note by the Secretariat Compilation of General Comments and General Recommendations Adopted by Human Rights Treaty Bodies' (1994) UN Doc HRI/GEN/1/Rev.1, para 10; and UN

certain groups – most notably, people living in poverty, women, and indigenous peoples[69] – this is an extremely important principle or guideline for policies and interventions.

The RBA rests on the state's legal obligation to respect, protect and fulfil the human rights of individuals.[70] Thus, for instance, this approach would view anti-poverty programmes as an obligation under international human rights law (stemming from the right to an adequate standard of living and the right to social security, for example),[71] rather than as a discretionary matter of charity or goodwill. Similarly, the component principles of the RBA – such as equality and participation – are themselves understood as rights, which generates a discourse and expectations of duties and obligations. RBAs are simultaneously an approach and an end in themselves; both the process and the outcome are equally important.[72]

## 4.2 RBA and Development

The processes and outcomes of development clearly overlap with and affect human rights, and indeed development is the field of practice where RBA has been most widely used. The link between development and human rights was formally enshrined in the Declaration on the Right to Development, adopted by the General Assembly in 1986,[73] although the right to development continues to prove highly contested. However, human rights has gained another, more secure footing in the development

---

Committee on Economic, Social and Cultural Rights, 'General Comment No. 20' (2009) UN Doc E/C.12/GC/20, paras 9 and 39.

[69] See for example, E Neumayer and T Plümper, 'The Gendered Nature of Natural Disasters: The Impact of Catastrophic Events on the Gender Gap in Life Expectancy, 1981–2002' (2007) 97/3 *Annals of the American Association of Geographers* 551.

[70] United Nations Development Programme (UNDP), *Rights Based Approach to Development Programming in the UNDP – Adding the Missing Link* (United Nations 2001); United Nations Population Fund (UNFPA), *A Human Rights-Based Approach to Programming: Practical Implementation Manual and Training Materials* (UNFPA and Harvard School of Public Health 2010).

[71] International Covenant on Economic, Social and Cultural Rights (adopted 16 December 1966, entered into force 3 January 1976) 993 UNTS 3 (ICESCR), arts 11 and 9 respectively.

[72] P Uvin, 'From the Right to Development to the Rights-Based Approach: How "Human Rights" Entered Development' (2007) 17/4 *Development in Practice* 597; UNFPA (n 70) 89.

[73] UN General Assembly, Declaration on the Right to Development: Resolution adopted by the General Assembly A/RES/41/128 (4 December 1986).

field since the 1980s, as development actors increasingly frame their work in human rights terms. Although its concrete impact and successes are hard to measure, the RBA to development is now widely used by bilateral development agencies, donors and international development NGOs such as Oxfam, ActionAid and CARE International. It is also used by the UN Development Programme (UNDP); indeed, human rights has, in theory at least, now been mainstreamed across the activities and programmes of all UN agencies.[74] A document of 'Common Understanding' on the human rights-based approach to development cooperation and programming was adopted by the UN Development Group in 2003,[75] and the approach is mandated for UN Country Teams.[76]

According to the development NGO CARE Denmark, RBA is a systemic approach: '[i]t changes the work of an organization by bringing it to reflect on and tackle not only immediate but also underlying causes of poverty'.[77] The RBA is contrasted with a needs-based approach: while a needs-based approach to development or poverty reduction might see vulnerability as being a symptom of poverty, a rights-based approach would view vulnerability as a structural issue and as an underlying cause of poverty. The former would see the poor as victims, who are then to be helped and accept assistance as and when it is provided, whereas the RBA would support them to participate in decision-making and to assert their rights.[78] 'The move from needs to rights, and from charity to duties, also implies an increased focus on accountability',[79] and thus the need for accessible accountability mechanisms and processes should be at the core of any rights-based approach to development.

### 4.3 Participation, Poverty and Power

The human rights-based understanding of participation is of great importance to environmental matters (as manifested to some extent in the Aarhus Convention). Participation was already a feature of much development practice prior to the spread of RBAs. During the proliferating

---

[74] See the UN Human Rights-Based Approach portal: http://hrbaportal.org/the-un-and-hrba (accessed 16 November 2014).

[75] UN Development Group (n 67).

[76] S McInerney-Lankford, 'Human Rights and Development: A Comment on Challenges and Opportunities from a Legal Perspective' (2009) 1/1 *Journal of Human Rights Practice* 51, 63.

[77] CARE Denmark, *Rights-Based Approach Guidelines* (CARE 2009) 2.

[78] Ibid 3.

[79] Uvin (n 72) 602–603.

neo-liberal reforms of the 1980s–90s, major international aid, finance and development institutions began to incorporate participation into technical approaches to development, seeing its value as a means of reducing costs and enhancing the efficacy of interventions. Through the neo-liberal lens, the subjects of development were largely seen as passive 'service users' or 'customers'.[80] However, partly due to the growth in civil society organisations, participation became increasingly contested, criticised as having been co-opted by dominant institutions and disconnected from social transformation.[81] However, once development actors began to frame their work in human rights terms, a re-politicisation of participation took place, with advocates of this approach envisaging 'transformative' participation that is both a means and an end and prioritises empowerment.[82] Now, the human rights approach to development clearly implies an 'absolute requirement of participation'.[83]

The right to participation in public, political, cultural, economic and social life is enshrined in a number of international human rights treaties, including the Universal Declaration of Human Rights (Articles 21 and 27) and the International Covenant on Civil and Political Rights (Article 25). Participation therefore holds a critical place in RBAs, as a necessary part of any rights-respecting policy or decision-making process and a right in itself. Consequently, RBAs are 'specifically concerned with finding ways of empowering those whose rights are denied to assess their condition, to identify the root causes of their marginalization and to take action – individually or collectively – to define, claim and realize their rights'.[84]

Participation is an integral part of empowerment, accountability and of ending cycles of deprivation and dependency in favour of the autonomy

---

80 A Cornwall and J Gaventa, 'From Users and Choosers to Makers and Shapers: Repositioning Participation in Social Policy', IDS Working Paper 127 (Institute of Development Studies 2001).

81 S Hickey and G Mohan (eds) *Participation: From Tyranny to Transformation* (Zed Books 2008).

82 S White, 'Depoliticising Development: The Uses and Abuses of Participation' (2006) 6/1 *Development in Practice* 142. White's 'ladder of participation' presents four forms of participation: (1) nominal, (2) instrumental, (3) representative and (4) transformative.

83 Uvin (n 72) 604.

84 A Ling, R McGee and J Gaventa, *Literature Review on Active Participation and Human Rights Research and Advocacy* (Institute of Development Studies 2010) 8.

and social inclusion of vulnerable groups.[85] Additionally, participation in decision-making processes also increases the chance that such policies will be effective, sustainable, inclusive and equitable: for example, there is little chance that poverty reduction programmes will work 'if there is no participation of the very poor in the interpretation and transformation of their reality'.[86] All these benefits of participation are clearly shown in the work of the NGO International Movement ATD Fourth World (ATD), which aims to find solutions to eradicate extreme poverty, working in partnership with people living in poverty, using a human rights-based approach. As well as developing material assets and human capital, the goal is to 'work towards building a consensus for a new social contract in which the extreme poor (and excluded people in general) assume the role of subjects in the process of transforming and recreating the institutions that regulate decision-making as well as the distribution and redistribution of assets'.[87] Clearly, the institutions and decision-making processes related to environmental governance and climate change could benefit from such a transformation.

A rights-based understanding of participation also pays increased attention to the way that power relations play out in certain contexts, recognising the fact that traditional 'participation' often tends to be *pro forma*, taking the form of information-giving or 'consultation' regarding decisions already made. Truly meaningful participation is dependent on other rights, such as non-discrimination and equality. Thus, under an RBA particular care should be taken to include and support those who are most excluded, including by providing accessible information and covering costs,[88] and taking into account inter- and intra-community power dynamics and socio-cultural practices.[89] A traditional approach to participation might involve asking whether women were included in a consultation, and the affirmative reply would cite a headcount of women present. A true human rights approach would interrogate the gender power dynamics more closely, including how many of the women present

---

85 For a fuller analysis, see M Sepúlveda, 'Report of the Special Rapporteur on Extreme Poverty and Human Rights' (2013) UN Doc A/HRC/23/36.

86 X Godinot and Q Wodon (eds) *Participatory Approaches to Attacking Extreme Poverty: Case Studies Led by the International Movement ATD Fourth World* (World Bank 2006) 8.

87 Ibid 9.

88 Uvin (n 72).

89 L Veneklasen et al., 'Rights-based Approaches and Beyond: Challenges of Linking Rights and Participation', IDS Working Paper 235 (Institute of Development Studies 2004).

actually spoke, and demand that both sexes are given the chance to represent their views, including, if necessary, through specially-targeted consultations and support.[90] The findings of a cross-country study conducted by the UK Inter-Agency Group on Human Rights Based Approaches,[91] show that RBA projects 'systematically increased the representation of poor and marginalized people', and although the non-RBA projects examined often emphasized securing participation, it was not done in a way that linked participation to inclusion and fulfilment of obligations.[92] The study also found that RBA projects featured more attempts to link citizens and state directly, rather than relying on an NGO mediating and liaising between them.[93]

## 4.4 Evidence for the RBA

The RBA is used in development for its potential to address and question structural and systemic causes of disempowerment and vulnerability to poverty. If it can succeed in this, is there good reason to believe it might be able to tackle the root causes of vulnerability to environmental harm? McInerney-Lankford of the World Bank argues that human rights should be systematically integrated into development policy and practice because rights are intrinsically valuable in protecting human dignity (which can be harmed by development interventions); and because they are instrumentally useful to enhance development processes, address certain types of social risk, ensure accountability and secure more equitable and sustainable development outcomes.[94] Certainly, environmental policy and practice could also benefit in these manners.

However, it must be acknowledged that there is not a solid, broad evidence base demonstrating the success (or failure) of rights-based approaches. This is partly due to the practical difficulties of measuring or evaluating human rights work, which generally aims at long-term social change and power shifts and cannot be evaluated by quantitative data

90 ActionAid, *People's Action in Practice: ActionAid's Human Rights Based Approach 2.0* (ActionAid International 2012) 59.

91 The Group includes the UK Department for International Development, HelpAge International, CARE UK, Save the Children UK and Oxfam.

92 UK Interagency Group on Human Rights Based Approaches, *The Impact of Rights-based Approaches to Development* (UK Interagency Group on Human Rights Based Approaches 2007) 44.

93 Ibid 39.

94 McInerney-Lankford (n 76) 52.

alone.[95] However, a World Health Organization (WHO) study based on experiences in four countries found that applying human rights to women and children's health policies and programmes did contribute to improved health outcomes for women and children. While the improvements found cannot be attributed exclusively to the RBA, 'there is plausible evidence that human rights contributed positively to health and health-related gains for women and children in the four countries, such as increased access to emergency obstetric care (Nepal), increased access to modern contraception (Brazil), reductions in early childhood mortality (Malawi) and increased vaccination coverage (Italy)'.[96] In addition, the findings produced by the UK Interagency Group on Human Rights Based Approaches:

> suggest that RBA projects are having considerably more success than non-RBA projects in attaining impacts that will lead to sustained positive change. RBAs tackle the underlying causes of poverty and disadvantage, and work in partnership with a wide range of stakeholders to address these causes. They link citizens and state in new ways and create systems and mechanisms that ensure that all actors can be part of accountable development processes.[97]

A 2007 study of RBA and non-RBA projects by CARE USA and Oxfam America does not conclude that one of the other has more or less impact.[98] Rather, both were found to produce different *types* of impact, with RBA projects addressing the underlying causes of poverty and exclusion rather than just the immediate and intermediate causes, producing less immediate and less tangible – but more sustainable – impacts.[99]

## 4.5 Conclusions on the RBA

RBAs can be interpreted and applied in a number of ways, and in some instances are justly subject to accusations of opportunism, misinterpretation or donor appeasement. Certainly, in the absence of a

---

[95] International Council on Human Rights Policy (ICHRP), *No Perfect Measure: Rethinking Evaluation and Assessment of Human Rights Work* (International Council on Human Rights Policy 2012).

[96] F Bustreo, P Hunt et al., *Women's and Children's Health: Evidence of Impact of Human Rights* (World Health Organization 2013) 13.

[97] UK Interagency Group (n 92) 7–8.

[98] CARE USA and Oxfam America, *Rights-based Approaches Learning Project* (CARE and Oxfam 2007).

[99] P Gready, 'Reasons to be Cautious About Evidence and Evaluation: Rights-based Approaches to Development and the Emerging Culture of Evaluation' (2009) 1/3 *Journal of Human Rights Practice* 380, 397.

grounding in people's daily needs and struggles for survival and dignity, rights-based approaches can be dismissed as merely a new technocratic fad that combines expert-driven social and economic interventions with legal change that may be irrelevant to people and communities.[100] Meanwhile, for human rights purists, especially lawyers, RBAs may feel like a dilution of the rights enshrined in legal standards.

This author takes the alternative view that 'the lack of consistency could be seen as a flowering of human rights, set free from the rather set ways of working which characterize the human rights mainstream'.[101] Despite potential shortcomings if used strategically, an RBA to climate change or the environment can be a useful tool to inform environmental efforts, especially those taking place at the local and regional level. A wider use of RBAs will help to ensure not only that responses to climate change and environmental degradation are more inclusive, appropriate and sustainable, but also that the process of – for example – climate change mitigation or adaptation is empowering, accountable, non-discriminatory and capacity-building in and of itself. This is a vital matter; as a UN guide to the RBA to climate change states: 'Inadequate mitigation and adaptation strategies can lead to human rights violations where adequate participation of local communities is not assured or if due process and access to justice is not respected for any necessary displacement'.[102]

Regarding RBAs to development, Gready argues that the 'deployment of rights by RBA specifically to address systemic, structural causes of poverty is ground-breaking; on the other hand, the challenges to power and marginalization at present appear localized'.[103] However, as RBAs become ever more widespread across development and environmental practice (undertaken by NGOs, UN agencies and bilateral aid agencies), their potential to challenge the status quo should grow.

---

[100] Veneklasen et al. (n 89) 3.

[101] Gready (n 99) 393.

[102] UN Office of the High Commissioner for Human Rights (OHCHR), *OHCHR Issue Note: Applying a Human Rights-Based Approach to Climate Change Negotiations, Policies and Measures* (United Nations 2010).

[103] Gready (n 99) 398.

## 5. A PRINCIPLED STRATEGY FOR PRACTICE

There are strategic merits for practitioners and activists in all of the three approaches discussed above, depending on need, tactics and circumstances. Crucially, environmental activists on the ground are already using human rights strategically, where they are felt to be appropriate or useful. Many are doing so without any prompting from or links with national or international human rights organisations, and many are combining human rights with other languages of social or climate justice. As Scholsberg observes, 'the diversity of concerns in environmental justice movements have been reflected in the broad and pluralistic definition of justice used';[104] it is therefore unsurprising that human rights have now been added to the mix.

This approach to human rights will be termed a 'principled strategic' one, although some would call it merely an 'instrumental' approach. Baxi writes that for the victims of the Bhopal catastrophe, human rights languages were not important in themselves but only as a means to the ends of justice.[105] This may be so, but the divide between the means and the end is not a simple one; the means can change the ends, create new ends, or be ends in themselves. This approach is strategic *and principled*, because, as Mark Goodale writes of his anthropological research in a Bolivian community legal services centre, 'the decision to embrace human rights discourse cannot be understood merely strategically, like making the decision to use genetically modified potato seed to increase crop yields. Rather, for many people in rural Bolivia, the shift in consciousness itself is the purpose of human rights discourse'.[106] Claims-making is a distinctive part of human rights practice and has intrinsic worth and radical potential.

The power of human rights lies not always in their legal forms, but in their conception of what it is to be human, their discursive power, in the claims-making itself. Human rights evolve and change through their use in grassroots movements, in local activism, in power struggles, applied to issues that the drafters of the Universal Declaration of Human Rights could not have envisaged. For Baxi, '[t]he message of Bhopal … constructs some new alternate futures beyond the new paradigm of trade-related, market friendly and environmentally hostile human

---

104 D Schlosberg, 'Theorising Environmental Justice: The Expanding Sphere of a Discourse' (2013) 22/1 *Environmental Politics* 37.

105 Baxi (n 28) 33–34.

106 M Goodale, 'Ethical Theory as Social Practice' (2006) 108(1) *American Anthropologist* 25, 31.

rights'.[107] This, surely, is a crucial outcome: movements which use human rights, even in a strategic/instrumental way, even when failing to achieve their desired ends (in this case, meaningful justice for Bhopal victims) can thus reconstruct human rights in more radical form, struggling free from the grip of the powerful. This is similar to the way in which women's rights activists have drawn attention to the limitation of traditional legalistic approaches to human rights work, going beyond the content of laws to reconceptualise rights as a process of political struggle by which people translate their needs and aspirations into demands and enforceable commitments by states.[108]

'Human rights' has achieved a remarkable degree of monopoly over conceptions and languages of justice, and its practitioners are often extremely protective over this status. However, in order for human rights to remain relevant, radical reformulations of human rights should not be rejected. Although they seem to rest on a somewhat static legal basis, the nature of human rights claims and duties is deeply political and constantly shifting, 'for what is socially and legally feasible today is never fixed, but a matter of political struggle'.[109] Nor should other languages of social justice be neglected. As Goodale states, 'the impact of human rights is never separate from the swirl of other sources of normative inspiration', be they community norms, state law, or religious standards.[110] After all, why should activists accept the limits placed on their demands, to ask for only what is already written, to demand only those human rights that emerged from previous struggles and have been consolidated into new forms of political power?[111]

Hoover and De Heredia contend that the dominant account of human rights is premised on 'an unsustainable division between moral philosophy and political action'. Rights, understood as inherently political and contested, are reconstructed in their account as a 'moralized politics'.[112] As an example of how individuals and groups 'reconstruct both the practice and idea of human rights in light of their own experience', they examine the discourse of the Zapatistas in Mexico and the Landless Peasant Movement (*Moveimento dos trabalhadores rurais Sem Terra*, or

---

107 Baxi (n 28) 44.

108 Veneklasen et al. (n 89).

109 Uvin (n 72) 603.

110 Goodale (n 106) 29.

111 J Hoover and MI De Heredia, 'Philosophers, Activists and Radicals: A Story of Human Rights and Other Scandals' (2011) 12 *Human Rights Review* 191, 215.

112 Ibid 193.

MST) in Brazil, two social movements not explicitly supported or recognised by mainstream human rights bodies or organisations, but which nevertheless use rights ideas and language in their claims and discourse. The MST, for example makes strategic use of the state framework of human rights to demand government accountability, security and redistribution and gain international appeal, while simultaneously occupying land and demanding radical changes to the economic order.[113] Here, the MST is one among many advocates and activists using 'universal human rights' to lend moral and political legitimacy to their work.

Thus, while conventional understandings of human rights make it difficult to alter the basic structures of sovereignty, representative government and economic decision-making, 'looking to the reality of social struggles, especially those not sanctioned by mainstream human rights bodies, provides important insight for reconstructing rights'.[114] As Žižek argues in his provocative article 'Against Human Rights', 'something that was originally an ideological edifice imposed by colonizers' can be 'taken over by their subjects as a means to articulate their "authentic" grievances' – like the Virgin of Guadalupe in newly colonised Mexico, whose appearance to an indigenous person allowed that population to appropriate Christianity 'as a means to symbolize their terrible plight'.[115] The struggle for environmental or climate justice has the potential to contribute to a much-needed task: the renewal and reconstruction of human rights in a more radical form.

The Universal Declaration of the Rights of Mother Earth,[116] drafted at the 2010 World Peoples' Conference on Climate Change and the Rights of Mother Earth in Cochabamba, Bolivia (Cochabamba Declaration) is one example of what Sally Engle Merry calls 'remaking human rights in the vernacular'.[117] In the Cochabamba Declaration, human rights are adapted and translated to a specific need: to a vision of human beings embedded in and dependent on nature, which includes recognition of the destructive effect of capitalism and affirms 'that to guarantee human rights it is necessary to recognize and defend the rights of Mother Earth'. The Declaration includes the 'inherent rights of Mother Earth and all

---

113 Ibid 212.

114 Ibid 205.

115 S Žižek, 'Against Human Rights' (2005) 34 *New Left Review*.

116 Available at http://www.rightsofmotherearth.com/declaration/ (accessed 16 November 2014).

117 SE Merry, 'Transnational Human Rights and Local Activism: Mapping the Middle' (2006) 108/1 *American Anthropologist* 38.

beings of which she is composed' (including the right to life, to maintain identify and integrity, and the right to clean air) and the obligations of human beings to Mother Earth. Why should this conception of rights be any less valuable, or universal, than those enshrined in the International Bill of Rights?

Of course, human rights practitioners should not force human rights tools and language on local activists.[118] As a recent study of activist organising in the United States shows, human rights may not always be the best choice of means or language, due to political and cultural obstacles to their acceptance.[119] However, even in 'hostile' contexts, many organisations inevitably find ways to remake rights in their particular vernacular. One US-based environmental activist organisation, the Community Environment Legal Defense Fund, works to establish 'community rights', in particular assisting communities to draft Community Bills of Rights which assert the 'right to local self-governance', so that 'communities are able to say "no" to [environmental] threats and "yes" to sustainability'. These Bills establish rights, including to clean air, to clean water and to sustainable energy, and prohibit 'activities that would violate these rights, such as fracking and GM seeds'.[120]

As Gearty states, '[s]uccessful social movements are absolutists in the pursuit of their interests but invariably open-minded about how they can be realized'.[121] Thus, those who are concerned with environmental protection and with human flourishing in a healthy environment should encourage, use or welcome a 'principled strategic' use of human rights, based on an analysis of what can best realise inevitably contextual visions of justice, dignity and flourishing.

---

[118] Although human rights education is desirable to ensure that the choice is at least available to those who may not be rights-aware.

[119] AC Finnegan, AP Saltsman and SK White 'Negotiating Politics and Culture: The Utility of Human Rights for Activist Organizing in the United States' (2010) 2/3 *Journal of Human Rights Practice* 307.

[120] Community Environment Legal Defense Fund (CELDF), 'Community Rights' (CELDF website, undated) http://www.celdf.org/section.php?id=423 (accessed 16 November 2014).

[121] C Gearty, 'Do Human Rights Help or Hinder Environmental Protection?' 2010 1/1 *Journal of Human Rights and the Environment* 7, 14.

## 6. NOTE: OBSTACLES IN INTERNATIONAL ENVIRONMENTAL GOVERNANCE

It is worth noting that recent years have seen human rights increasingly sidelined in international environmental forums, in contrast to the growing prominence of environmental matters in human rights discourse. For example, while the outcome agreement of the Cancún COP (2011)[122] included substantive references to human rights, they were notably absent from the decisions and documents negotiated since at Durban, Doha and Warsaw.[123] This marginalisation of human rights is problematic. No governmental or inter-governmental actor seriously questions the interdependence of human rights and environmental quality, and environmental issues are now taken seriously in human rights fora, but yet the international environmental structures and institutions (including the UNFCCC) go about their business with barely even a tip of the hat to human rights. This lack of policy coherence and the reality of institutional fragmentation at the international level is even more visible when the WTO, the IMF and the World Bank are included in the analysis. The environment, like the issue of global poverty, is an area where most governments speak in one voice in one arena and a different one entirely in another – promising to honour commitments to fight poverty or climate change at home and abroad, while simultaneously making trade or investment deals that cannot fail to undermine these commitments.

## 7. CONCLUSION

This analysis has shown that human rights practice, even in its dominant forms, is not monolithic. Hence, it can offer various avenues for raising and addressing environmental concerns: (1) using human rights law and its mechanisms in as progressive a way as possible, pushing for recognition of rights claims related to the environment; (2) the use of RBA on the ground by development and environmental protection practitioners;

---

[122] UNFCCC Conference of the Parties, Decision 1/CP.16 The Cancun Agreements: Outcome of the work of the Ad Hoc Working Group on Long-term Cooperative Action under the Convention. FCCC/CP/2010/7/Add.1 (2010).

[123] UNFCCC Conference of the Parties, Decisions adopted by the Conference of the Parties FCCC/CP/2011/9/Add.1 (2011); UNFCCC Conference of the Parties, Decisions adopted by the Conference of the Parties FCCC/CP/2012/8/Add.1 (2012); UNFCCC Conference of the Parties, Decisions adopted by the Conference of the Parties FCCC/CP/2013/10/Add.1 (2013).

and (3) supporting alternative and radical conceptions of rights and environment claims emanating from grassroots contexts. Each of these can be of use, if undertaken in a radical and/or strategic way, recognising the limitations and strengths of the current human rights regime, while seeking to improve it.

However, further action beyond these approaches is clearly necessary. For example, environmental and human rights campaigners must maintain their challenge to the insidious notion of corporate humanity and their efforts to hold corporations and International Financial Institutions accountable for their negative impacts on human rights. Other languages and conceptions of justice are also crucially important (that of distributive justice, for instance).

A multi-faceted approach is crucial because human rights practice tends to tackle environmental issues at a (comparatively) micro-level, at least in the short-term: an RBA project promoting participation of marginalised people in environmental decision-making in one community; a grassroots activist group using radical rights discourse to protest river pollution in another; a court ruling that a rights violation has been committed against an individual exposed to environmental toxins. Undoubtedly, such actions are radically important (and even subversive), especially in combination. However, in order to achieve truly sustainable environmental protection, a profound shift in dominant political-economic ideologies will be necessary.

Ron Dudai argues that although human rights can attempt to ensure the fairness and appropriateness of *responses* to climate change, it does not have a role in engaging with the *substance* of climate change policy.[124] This is true in some respects: clearly climate change and broader environmental issues often require technical and scientific approaches and solutions. However, human rights can and should inform environmental policies, actions and agreements at the international and national level, including in such 'technical' areas as technology transfer,[125] emissions targets, REDD[126] and biofuels. These are, after all, issues that engage profound questions of justice, and could promote or obstruct the

---

[124] R Dudai, 'Climate Change and Human Rights Practice: Observations On and Around the Report of the Office of the High Commissioner for Human Rights on the Relationship between Climate Change and Human Rights' (2009) 1/2 *Journal of Human Rights Practice* 294.

[125] See International Council on Human Rights Policy, *Beyond Technology Transfer: Protecting Human Rights in a Climate-constrained World* (International Council on Human Rights Policy 2011).

[126] Reducing Emissions from Deforestation and Degradation.

enjoyment of human rights on the ground. As Gearty argues, the human rights approach is an 'essential ethical component of a proper response to climate change' because it can help to ensure that decisions in this area do not again result in the world's poor being made to pay a disproportionate price.[127] Human rights offer a strong ethical and legal basis from which to approach these issues, and can help direct resources, technologies and/or action where they are most urgently needed, keeping a focus on the most vulnerable and marginalised communities.

Environmental protection, like poverty reduction, requires a persistent engagement with structural causes, and a coordinated, multifaceted approach at all levels of global and national governance. If human rights is fit for its role as the dominant *lingua franca* of global social justice, it must have something to contribute. Human rights practice has evolved in various directions in order to deal with the growing threat of environmental degradation and climate change; it can and should play a major part in response to these existential threats to human flourishing. It has been argued here that there is no one 'correct' approach, based on an understanding that human rights practice should be principled yet strategic, profoundly context-dependent, and ultimately directed at empowerment and shifting power hierarchies. This may seem unfocused and even a dilution of legally recognised human rights. However, while human rights *practice* can be an end in itself – creating popular participation, empowerment, socio-cultural change – human rights themselves, as law or concept, are worthless if they are not useful to social justice struggles on the ground. Therefore, human rights practitioners must remain open-minded and imaginative regarding efforts to save and protect Earth's environment. Ultimately, this is a battle that may well define the future of human rights, which were – and continue to be – developed, discarded and reconstructed within specific contexts and political struggles.

---

127 Gearty (n 121) 21.

# 8. Schopenhauer's *Mitleid*, environmental outrage and human rights

***Tom Kerns****

## 1. INTRODUCTION

When we feel a sense of outrage on witnessing great wrongs perpetrated against persons or communities, where does that outrage come from? What accounts for it and is it justifiable? The thesis of this chapter is that Schopenhauer's ethic of *Mitleid* (compassion) both explains where this sense of outrage comes from and why it is justified; and further, that widely accepted human rights standards confirm the validity of that experience of outrage and the felt need to act on it.

The modern human rights movement began in outrage, and the modern environmental movement found its human rights footing only when it too began to feel outrage at the increasing number, complexity and scope of environmental abuses that impacted human persons. The Universal Declaration of Human Rights (UDHR), a founding document in the modern human rights movement, reminds us in the second recital of its Preamble of the recent horrific events in World War II, and of those 'barbarous acts which have outraged the conscience of mankind'.

The Preamble also reminds states that it is in their self-interest to promote human rights,[1] an appeal to the motive that Schopenhauer will term 'egoism', but emphasis in the Preamble is placed on reminding

* This chapter was first published as T Kerns, 'Schopenhauer's *Mitleid*, Environmental Outrage and Human Rights' (2013) 5/3 *Oñati Socio-Legal Series* 931–52.

[1] Recitals three and four of the UDHR Preamble refer to states' self-interest in promoting human rights: 'Whereas it is essential, if man is not to be compelled to have recourse, as a last resort, to rebellion against tyranny and oppression, that human rights should be protected by the rule of law', 'Whereas it is essential to promote the development of friendly relations between nations'.

states of the moral outrage that the human community felt and expressed at governments whose behaviours showed such disregard for human suffering.

The injuries and devastating suffering that so often result from exposure to environmental toxics, from damage to the environment and from climate change often evoke a similar sense of moral outrage. This chapter explores the grounds and implications of that outrage.

## 2. OUTRAGE EXPERIENCES

Most of us can probably think of environmental situations that have impacted people in serious ways, about which we have felt appalled and horrified. Perhaps it was aerial pesticide sprays of large populated areas that resulted in devastating health and economic impacts on residents; or heavy industrial pollution of water and air that had dramatic impacts on nearby residents; or oil and gas drilling or refining operations that have affected families and communities and contributed to climate change.

Imagine the family, for example, whose next door neighbours have leased their land to an oil and gas company, and a six-acre drill pad has now been erected on the neighbour's land just 200 yards away from this family's home. Imagine that shortly after drilling began, the family's well water suddenly developed a pronounced chemical taste and they no longer felt safe drinking, cooking or showering with it. Imagine that the children began experiencing respiratory and other health problems within a few days of the start of drilling operations, so that the family sees no other option than to move away and live elsewhere – only to discover that real estate values in their neighbourhood have plummeted because of the presence of hydraulic fracturing, so they can no longer sell their home, and their insurance covers none of their losses because losses associated with fracking have been exempted.

Or consider the experience of the young Latina student in my class one quarter who suffered from a variety of health challenges – respiratory, neurological, skeletal, developmental, reproductive, etc. – most of which she had endured since birth. At one point in the class when I was lecturing on environmental toxics, pesticides and health impacts, you could almost see a light go off in her head. She raised her hand and asked whether I thought it might be possible that her mother's exposure to pesticides in the agricultural fields of Eastern Washington could be a factor in her own health situation. 'When my mother was pregnant with me her job was to be a spotter', she explained. 'When the pesticide spray plane flew low along the rows of berries, my mother's job was to stand at the end of each long row and wave her hands in the air as the spray plane passed over to signal the pilot that he should turn around at that point and fly back down the next row. She did that all day long when the crop dusters were spraying. Do you think that could have been a factor in my health?'[2] It is not the environment, of course, that is responsible for injuring people in these situations; it is the individuals, corporations or governments that have done things, or allowed things to be done, to the environment that have then resulted in injury to people.

If we feel outrage at situations like these, this chapter asks what accounts for that outrage, where exactly it comes from and what it is grounded in. The chapter argues, first, that the philosophy of Arthur Schopenhauer provides an ethical and metaphysical account of where such moral outrage comes from, how it comes to be and why it is so powerful, and secondly, that widely accepted moral norms expressed in human rights documents then confirm the validity of those feelings of outrage. Both these insights have implications for activist responses to issues at the nexus between human rights and the environment.

The philosophical claim to be made in this chapter is that Schopenhauer's ethic of compassion, his *Mitleids-Moral*,[3] provides a better account of our experience of moral outrage than do other ethical systems such as Aristotle's ethic of virtue and happiness, Immanuel Kant's categorical imperative or John Stuart Mill's consequentialist, greatest happiness for the greatest number, principle, or even ethical systems based on religious beliefs about rewards and punishments in an afterlife. These philosophies and theologies certainly do have their strengths, but accounting for the experience of moral outrage is not one of them.

This chapter uses the terms 'outrage' and 'moral outrage' in their ordinary sense of an unusually strong reaction of anger, shock or indignation – also described as feeling appalled, horrified, aghast or disgusted – particularly at witnessing or learning of great wrongs done to others. If we followed Plato's psychology, we would attribute this feeling to the part of the soul he calls the spirited part,[4] the part that loves justice and gets angry at injustice.

---

2 The horrific Union Carbide disaster in Bhopal, India, in 1984 could also serve as a trenchant example of situations that evoke moral outrage. See U Baxi, 'Writing about Impunity and Environment: The "Silver Jubilee" of the Bhopal Catastrophe' (2010) 1 *Journal of Human Rights and the Environment* 23–44.

3 DE Cartwright, 'Compassion and Solidarity with Sufferers: The Metaphysics of Mitleid' (2008) 308 *Eur J Philos* 292–310.

4 Described in Socrates' second speech in Plato's *The Phaedrus*.

As Lynn Hunt says in *Inventing Human Rights: A History*, 'we are most certain that a human right is at issue when we feel horrified by its violation'.[5] When we see or hear of such horrible injustices we are often seized with shock and outrage and blurt out, if only to ourselves: How awful! How could anyone do such a thing? When we ask that question of ourselves, though, what exactly is it we are asking? Are we asking how someone could have so little fear of punishments in the afterlife? Probably not.[6] Are we asking how anyone could care so little about being a virtuous person (as in Aristotle's ethic)? Again, probably not. Are we asking how anyone could so completely overlook their duty to act on universal principles of moral obligation (Kant's ethic)? Almost surely not. Are we asking how anyone could so inaccurately estimate the sum of beneficial and harmful consequences (Mill's ethic)? No. Schopenhauer thinks the sense of our question is something closer to: How is it possible for anyone to be so utterly bereft of human compassion that they could do, or allow to be done, such a heartless deed.

## 3. SCHOPENHAUER'S 'HUMBLE PATH TO ETHICS'[7]

The primary source for understanding Arthur Schopenhauer's (1788–1860)[8] ethical philosophy is *On the Basis of Morality*,[9] an essay he submitted in 1839 to the Royal Danish Society of Scientific Studies for its prize essay contest about the foundations of morality. His essay was the contest's only submission. It directly addressed and answered the question but was not awarded the prize, probably because the judges were irked by his acerbic references to some highly respected German philosophers of the day, particularly Hegel, whom he considered to be bombastic, wordy, shallow and an empty 'philosophaster'. The Academy's one-paragraph formal judgement about Schopenhauer's essay

---

[5] L Hunt, *Inventing Human Rights: A History* (New York, Norton 2008) at 26. Her use of 'horrified by' will be taken as being equivalent to 'outraged by'.

[6] We will see later that I am paraphrasing Schopenhauer here.

[7] The term is from G Mannion, Schopenhauer, Religion and Morality: The Humble Path to Ethics (Aldershot, Ashgate 2003) 232.

[8] Born 22 February 1788 in Danzig (Gdańsk), died 72 years later in 1860 in Frankfurt.

[9] The two English translations used for this paper are: A Schopenhauer, *On the Basis of Morality*, Translated by EFJ Payne, Introduction by DE Cartwright (Indianapolis, Hackett 1995), and A Schopenhauer, *On the Basis of Morality*, Translated by DE Cartwright and EE Erdmann, Introduction by C Janaway (Oxford, OUP 2010).

claims that the essay did not directly answer the question posed, but the judges' concluding sentence says it all: 'Finally, we cannot pass over in silence the fact that several distinguished philosophers of recent times are mentioned in a manner so unseemly as to cause just and grave offense'.[10]

Schopenhauer scholar David Cartwright offers quite a different assessment, one less tainted by the tender sensitivities of the Danish Academy's panel of judges. 'Arthur Schopenhauer deserves to be considered a first-rate moral philosopher', he says,

> because of his analysis of the ethical significance of compassion (*Mitleid*). Although his ethics contains other important insights, it is his multi-faceted analysis of compassion which is its crowning jewel. The depth of Schopenhauer's understanding of the nature and ethical importance of this emotion is unparalleled in the history of Western philosophy.[11]

Although Schopenhauer himself was no saint (witness his misogyny, for example, and a brief review of his biography[12] would turn up more), he did not believe that a person's faults disqualified them from writing about ethics.[13] 'It is just as little necessary', he says, 'for the saint to be a philosopher as for the philosopher to be a saint',

> just as it is not necessary for a perfectly beautiful person to be a great sculptor, or for a great sculptor to be himself a beautiful person. In general, it is a strange demand on a moralist that he should commend no other virtue than that which he himself possesses.[14]

## 4. PURPOSE OF ETHICS

Schopenhauer begins by asking what the central job of ethics is, what it intends to study and what its methods are. Philosophers have given quite different answers to these questions, of course, most involving analyses and reasoned justifications for 'ought' and 'ought not' statements. Schopenhauer, though, sees ethics not as a set of justifications for

---

10 Schopenhauer (1995) (n 9) 216.

11 Quoted in G Mannion, *Schopenhauer, Religion and Morality: The Humble Path to Ethics* (Aldershot, Ashgate 2003) 198.

12 The best available is the welcome new biography by Cartwright, 2010: DE Cartwright, *Schopenhauer: A Biography* (Cambridge, CUP 2010).

13 MA Fox, '"Boundless Compassion": The Contemporary Relevance of Schopenhauer's Ethics' (2006) *Eur Leg* 37–71.

14 A Schopenhauer, *The World as Will and Representation, Volume* I. Translated by EFJ Payne (New York, Dover Publications 1969) 383.

prescriptive claims about how people *ought* to behave, but instead as a purely empirical study of how people actually do behave, with an eye to the moral dimension of their actions, along with an explanation, both psychological and metaphysical, for why they behave that way. The purpose of ethics, he says,

> is to indicate, explain and trace to its ultimate ground the extremely varied behaviour of [people] from a moral point of view. Therefore there is no other way for discovering the foundation of ethics than the empirical, namely, to investigate whether there are generally any actions [at all] to which we must attribute *genuine moral worth.*[15]

Schopenhauer's answer to that question will be yes, there are acts of genuine moral worth, justice and loving-kindness, but that our accounting for them will entail both explanation and, at bottom, a genuine mystery. The explanations, he believes, must be empirically based, not spun from philosophically abstract concepts in the academies. Explanations will need to be clear, intuitively evident and understandable by everyone, whether learned or not and whether they are abstract thinkers or not. No abstract, artificial combinations of abstruse concepts will be acceptable.

## 5. CRITERION FOR ACTIONS OF MORAL WORTH

To begin with, the non-moral incentives to action (for Schopenhauer) include egoism and malice. Malice, in which the incentive to action is to cause suffering in another, is morally reprehensible and is the actual and conceptual opposite of morally noble actions. Egoism, i.e., acting in one's own self-interest, which is by far the most common motivation for human actions and which shows itself in almost limitless forms, is, in his view, morally neutral.

The only actions that would have moral worth[16] are those done for the benefit of someone else, someone who is not me. A morally worthy act must therefore exclude actions motivated by self-interest or by malice.

> If a self-interested motive is the only one, its discovery entirely destroys the moral worth of an action; and if such a motive acts as an accessory, the moral worth of the action is reduced by its discovery. The absence of all egoistic motivation is therefore, *the criterion of an action of moral worth.*[17]

---

[15] Schopenhauer (1995) (n 9) 130.

[16] Ibid 143.

[17] Ibid 139–40, Schopenhauer's emphasis.

In Schopenhauer's view, egoism and moral worth absolutely exclude each other. As long as egoistic motivation is present – even if the anticipated self-benefit at which the action aims is remote, perhaps even in an after-life, and even if the goal aimed at is to improve my character or to 'perfect myself – as long as egoistic motivation is present, the action resulting from it does not have moral worth: 'If an action has as its motive an egoistic aim, it cannot have any moral worth. If it is to have moral worth, its motive cannot be an egoistic aim, direct or indirect, near or remote'.[18] There is thus only one case in which an act has moral worth and that is when the act is done solely to benefit another or to prevent or relieve the suffering of another.

It is worth asking at the outset whether any such actions exist, i.e., whether any actions 'exclusively centred in the weal and woe of someone else'[19] exist, or whether all human acts are egoistic, done for some purpose that is intended to benefit me. In other words is the class of morally worthy acts a populated or an empty class? Schopenhauer believes there are indeed acts done purely to benefit another or to prevent or ease their suffering. The key challenge, though, is to 'find out *what it is* that can move a [person] to actions of this kind'.[20] How is it possible, in other words, to be moved to act solely for another's well-being: 'How is it possible for *another's* weal and woe to move my will immediately, that is to say, in exactly the same way in which it is usually moved only by my own weal and woe?'[21] That is the question.

Schopenhauer's answer is 'compassion' ('*mitleid*' from '*leid*', suffering), a concept that conveys the sense of *suffering with* another. Acting to prevent or to alleviate another's suffering, just as I might otherwise act to prevent or to alleviate my own, entails my understanding, or even experiencing, the other's suffering as I normally experience only my own. 'But this [says Schopenhauer] necessarily presupposes that, in the case of *his* woe as such, I suffer directly with him',

> I feel his woe just as I ordinarily feel only my own; and, likewise, I directly desire his weal in the same way I otherwise desire only my own. But this requires that I am in some way *identified with him*, in other words, that this

---

[18] Ibid 141.

[19] Ibid 143.

[20] Ibid 140, Schopenhauer's emphasis.

[21] Ibid 143, Schopenhauer's emphasis. And he phrases the question again at 165: '[H]ow is it possible for a suffering which is not *mine* and does not touch *me* to become just as directly a motive as only my own normally does, and to move me to action?'

> entire *difference* between me and everyone else, which is the very basis of my egoism, is eliminated, to a certain extent at least.[22]

For Schopenhauer the compassionate actor feels, in some mysterious way, identified with the other person, feels their suffering in such a way that the difference between that person and himself is eliminated or at least reduced: 'As soon as this compassion is aroused, the weal and woe of another are nearest to my heart in exactly the same way, although not always in the same degree, as otherwise only my own are. Hence the difference between him and me is no longer absolute'.[23] Compassion is thus a kind of participation in the other's sufferings: 'We suffer *with* him and hence *in* him; we feel his pain as *his*, and do not imagine that it is ours'.[24] This, according to Schopenhauer, is the great mystery of ethics, and is astonishing. Compassion, for him, is *the* primary ethical phenomenon and yet presents itself to us as a mystery, i.e., as not entirely explainable. 'This event is certainly astonishing, indeed, mysterious. In fact it is the great mystery of ethics; it is the primary and original phenomenon of ethics, the boundary mark beyond which only metaphysical speculation can venture to step.'[25] (The metaphysical explanation will be discussed below.)

Since it is only weal (well-being) and woe (suffering), whether mine or another's, that moves the human will to act, there are in principle only four possible classes of motivation for human actions: (1) acting to promote my own weal and interests, which Schopenhauer terms 'egoism'; (2) acting to bring about my own woe and suffering, which Schopenhauer does not name and does not discuss;[26] (3) acting to bring about suffering and woe in another person, which he terms 'malice', and is at the far negative end of moral value; and (4) acting to prevent or ease another's suffering or on behalf of their well-being, which he terms

---

22 Ibid 143–44, Schopenhauer's emphasis.

23 Ibid 144. '[T]he process here analyzed is not one that is imagined or invented; on the contrary, it is perfectly real and indeed by no means infrequent. It is the everyday phenomenon of compassion, of the immediate participation … primarily in the sufferings of another, and thus in the prevention or elimination of it [sic]' (ibid).

24 Ibid 147.

25 Ibid 144.

26 Schopenhauer does not discuss the second category, acting to bring about my own woe. Is it an empty category or are there behaviours that would fit it? Masochism perhaps? Spiritual asceticism? In a letter to Johann August Becker, 10 December 1844, he claimed that the desire for one's own woe possessed 'ascetic value'. See (n 3).

'compassion'. In actuality, though, Schopenhauer focuses his discussion primarily on only two of these motivations: egoism, which he says accounts for the vast bulk of human actions, and compassion. And further, Schopenhauer argues that it is only sympathy with another's *suffering*, not with their well-being, that stirs us to compassion: 'Direct sympathy with another is restricted to his *suffering*. It is not roused, at any rate not directly, by his *well-being*; on the contrary, in and by itself this leaves us unmoved'.[27] And the reason for this, explains Schopenhauer, is that pain and suffering are the positive, substantial reality. Satisfaction, happiness and contentment are negative only, in that they are simply the absence of suffering.

> The reason for this is that pain, suffering that includes all want, privation, need, in fact every wish or desire, is *that which is positive and directly felt and experienced*. ... [T]he nature of satisfaction, enjoyment, and happiness consists solely in the removal of a privation, the stilling of a pain ...[28]

Suffering, in other words, is the primary ethical datum, and our concern for suffering, when it occurs in another person, is what moves us to act compassionately.[29] We can act compassionately in two ways, either by avoiding causing or increasing another's suffering, or by undertaking an action to relieve their suffering.

## 6. THE VIRTUE OF JUSTICE

Justice is the name of the virtue by which a person refrains from doing injury to another, and is the first of Schopenhauer's two cardinal virtues, loving-kindness being the other.

Imagine that your pursuit of a given egoistic aim such as comfort, pleasure or profit were to entail, as a side effect, causing injury or harm

---

27 Schopenhauer (1995) (n 9) 145, Schopenhauer's emphasis.

28 Ibid 146, Schopenhauer's emphasis.

29 A more recent expression of the importance of compassion can be found in the words of former Anglican Bishop Christopher Senyonjo of Uganda who appeared in *Call Me Kuchu*, a film about the repression of gay people in his country. He said that lack of compassion is at the root of human rights abuses: 'When people don't regard others as real human beings, they have no compassion for other people. They do anything they want to them, as if they were not human beings'. Voice of America, http://www.voanews.com/khmer-english/news/Movies-About-Human-Rights-Screened-in-New-York–161199605.html (accessed 10 October 2014).

to another person in the process. It would be the virtue of justice, stemming from compassion, that would prevent you from pursuing that aim (or at least from pursuing it in a manner that causes woe to the other) and that restraint would be morally worthy. You would be acting justly.

> [T]he first degree of the effect of compassion is that it opposes and impedes those sufferings which I intend to cause to others by my inherent anti-moral forces. It calls out to me 'Stop!'; It stands before the other [person] like a bulwark, protecting him from the injury that my egoism or malice would otherwise urge me to do.[30]

Applying this analysis, if a corporation, for example, acting in pursuit of its own egoistic (economic) ends, were to injure or to harm others in the process, that would be a violation of the virtue of justice and thus morally blameworthy.

If the virtue of justice were to be expressed in a maxim, that maxim would be *Neminem laede*, 'Harm no one'. Refrain from causing harm to others. Maxims and norms of this sort (perhaps including human rights norms) are important for living a moral life. Maxims are the fixed forms in which wc store our sense of compassion. They are the 'receptacle or reservoir which stores the habit of mind'.[31] So if a person is following the maxim of justice, to harm no one, compassion still acts indirectly in them when they are respecting the norms of justice. Even if this person acts primarily out of a sense of duty and respect for the norms of justice, compassion still (says Schopenhauer) remains always ready to come forward if the 'established maxim' is not working. 'Nothing', says Schopenhauer, 'will bring us back to the path of justice so readily as the mental picture of the trouble, grief, and lamentation' of the sufferer.[32] This argument underscores the well-recognised importance of personal narratives in environmental human rights work, because such narratives make immediately clear the sufferings of others and evoke the public compassion that can lead to change.

What Schopenhauer has underscored here is that injustice and moral wrong consist in causing or adding to the sufferings of another. And again it is 'wrong' that forms the positive concept here, while morally worthy acts are those that lead to the cessation, easing or prevention of suffering. In support of this claim of the conceptual primacy of injury and wrong, Schopenhauer quotes Hugo Grotius: 'Justice is that which is

---

30 Schopenhauer (1995) (n 9) 149.
31 Ibid 150.
32 Ibid 152.

not unjust'.[33] This may initially sound tautologous but Schopenhauer intends it to underscore his claim that injustice is the primary concept from which the concept of justice is derived.[34]

This formulation is reminiscent of Alan Dershowitz's argument in *Rights from Wrongs*[35] that rights emerge out of the experience of great wrongs, and that as new wrongs become possible and actual – pollution of the atmosphere, toxification of potable water, etc. – so new rights become evident, such as the rights to clean air and water. New rights emerge, in other words, in response to new ways in which people can be injured. This view also allows for the existence of purely moral rights that are antecedent to rights in law, because this explanation is based on the experience of actual empirical injury. The concept of rights as moral norms would thus for Schopenhauer precede the emergence of rights based in law.

> The concepts *wrong* and *right* are synonymous with doing harm and not doing harm, and to the latter belongs also the warding off of injury. They are obviously independent of, and antecedent to, all positive legislation. Hence there is a purely ethical or natural right … that is independent of all positive statute.[36]

Thus on the question as to whether rights exist only *after* they have been defined in law or exist as moral rights prior to their implementation in law, Schopenhauer comes down clearly on the side of the latter.[37]

---

33 H Grotius, *The Law of War and Peace* (I, 1,3).

34 On the question of which notion, right or wrong, is conceptually prior, Schopenhauer also cites Dante who, when asked 'Who knows what is good?' responds '[h]e who knows what is bad': quoted in A Schopenhauer, *On the Will in Nature* (Oak Grove Kindle 2011), Kindle location 453–55.

35 A Dershowitz, *Rights from Wrongs: A Secular Theory of the Origins of Rights* (New York, Basic Books 2005) 7–8.

36 A Schopenhauer, *On the Basis of Morality* (1995) (n 9) 154, Schopenhauer's emphasis.

37 Johannes Morsink refers to the view that rights exist only after they have been expressed in law as the fallacy of implementation: 'This fallacy does to human rights what the functionalism and behaviorism in psychology did or still do to our mental states. These schools of thought in psychology translate mental states out of existence as irrelevant to the real business of science. Just as some doctors and neurologists only want to know about the bodily behavior and neurobiological condition of their patients, so some human rights theorists are only interested in the instruments of international human rights law. And just as their behaviorist counterparts tend, for all practical purposes, to ignore people's mental states, so purists in the legal human rights field ignore the metaphysics of

Even more egregious than harming another is what Schopenhauer terms 'double wrongs', or 'treachery'. Double wrongs occur when a person whose job is to protect someone both (a) fails to protect and (b) actually harms the person he is supposed to protect.

> This *double injustice* occurs when anyone has expressly undertaken the obligation to protect someone else in a definite respect; consequently the non-fulfillment of the obligation would in itself be an injury to the other person, and thus a wrong; but in addition, he now attacks and injures the other [person] at the very spot where he should protect him. Such is the case, for example, when the appointed watchman or attendant becomes a murderer, the trusted custodian becomes a thief, the guardian defrauds his ward of her property, the lawyer prevaricates, the judge allows himself to be bribed, and the man asked for advice deliberately gives some dangerous and pernicious counsel.[38]

Examples of this in today's news are too numerous to mention: the investment manager who commits fraud for personal gain, the football coach who abuses children under his care, the bishop who protects abusing priests instead of the children, etc., and we could all probably think of similar examples committed both by individuals and institutions.

## 7. THE VIRTUE OF LOVING-KINDNESS

If justice, the first cardinal virtue, consists in refraining from causing injury to another, the second cardinal virtue, loving-kindness (philanthropy, *caritas*, *agape*), consists in and taking action positively to relieve or to prevent another's suffering. If justice is the virtue of not injuring and its maxim is *Neminem laede* ('Harm no one'), loving-kindness is the virtue of actively helping and its maxim is *Omnes, quantum potes, juva*: 'Help everyone as much as you can'. Acts of loving-kindness may be easy or difficult and may sometimes entail sacrificing one's own health, freedom, property or life to prevent or alleviate another's need or distress. Schopenhauer, not a great lover of religions, does credit Christianity for its emphasis on the virtue of loving-kindness and refers to compassion as

---

inherence because they are only interested in how the international legal system works on a practical level' (J Morsink, *Inherent Human Rights: Philosophical Roots of the Universal Declaration* (Pennsylvania, University of Pennsylvania Press 2009) 47).

[38] Schopenhauer (1995) (n 9) 156, Schopenhauer's emphasis.

'Christianity's greatest merit'.[39] He also notes, however, lest Christianity be given too much credit, that this virtue was being extolled in Asia 1000 years earlier in the Vedas and Buddhism. Loving-kindness is thus the second cardinal virtue and from it flow all the 'duties of virtue' and love, i.e., not just the moral minimums of justice.

So both cardinal virtues, justice and loving-kindness, stem from compassion, which is a kind of 'participation in' another's suffering. How that 'participation' happens, though, is a genuine mystery:

> But how is it possible for a suffering that is not *mine* and does not touch *me* to become just as directly a motive as only my own normally does, and to move me to action? … I … *feel* [the other person's suffering] *with him, feel it as my own*, and yet not *within me*, but *in another person.*[40]

This is, for Schopenhauer, *the* mystery of ethics. 'And yet it happens every day [he says]; everyone has often experienced it within himself; even to the most hard-hearted and selfish it is not unknown'.

> Every day it comes before our eyes, in single acts on a small scale, wherever, on the spur of the moment and without much reflection, one man helps another, hastens to the assistance of one whom he has seen for the first time, and in fact sometimes exposes even his own life to the most obvious danger for the sake of that [person], without thinking of anything except that he sees the other's great distress and danger.[41]

And yet how could this be possible? What is it about the underlying metaphysical substructure of the world that allows for this to happen?

## 8. METAPHYSICAL GROUNDING

This 'participation in' another's suffering is, says Schopenhauer, a genuine mystery. How does it happen? What is its undergirding? How does it relate to the underlying nature of things? These questions require some attempt at explanation. Thus, what has up until now been the basis for explaining morally worthy acts (compassion), now becomes the problem.[42] How is it to be explained? To answer this question Schopenhauer must now explore the metaphysical underpinnings of compassion –

---

39 Ibid 163.

40 Ibid 165, Schopenhauer's emphasis.

41 Ibid 166.

42 Ibid 203.

at which point he has to leave the realm of the empirical and phenomenal, and enter a realm in which analysis, suggestion and analogy are the most useful tools.

Schopenhauer's view of the ultimate nature of things takes up where Immanuel Kant left off in his *Critique of Pure Reason*. He accepts Kant's fundamental distinction between the world of appearances (phenomena) and the underlying true reality (noumena), and also accepts Kant's notion of the basic categories of perception, such as time, space and plurality, which are our contributions to perception and which shape and condition our entire experienced world.

For Schopenhauer, the phenomenal world, the world of appearance, corresponds to Hinduism's concept of Maya, the world of illusion, and to what the Greek philosophers termed the world of becoming, as opposed to the world of being. It is the world of our experience, and as it comes streaming in through our senses it gets shaped into a perceived world that is conditioned by time, space, plurality and the other categories of perception. The noumenal world, according to Schopenhauer, i.e., the true world of being that exists beneath these ephemeral appearances, is not shaped by space, time and plurality, but exists outside or beyond them.

Schopenhauer's term for the fundamental noumenal reality that underlies all the phenomenal appearances is *Der Wille*, a concept that has little or nothing to do with our usual concepts of human choosing, willing and acting. I tell students that if they think of *Der Wille* as 'The Force' in *Star Wars*, they won't be too far off. *Der Wille* is the underlying energy, elan, drive, or Will to Be that manifests itself in all the various particular forms we experience in the world of sensory multiplicity spread out in perceptual space and time.

All these multiple phenomenal beings, i.e., all the temporal-spatial multiplicities of our experience, are, for Schopenhauer, phenomenal expressions of the one underlying noumenal reality, *Der Wille*. All the apparently separate individuals in our experienced world only *appear* to be separate individuals. At root they, we, are all just different expressions, different manifestations, of the same fundamental underlying noumenal reality. This fundamental underlying identity of all individual persons, of all living things, and in fact of all things, is central for Schopenhauer, and is his explanation, at the metaphysical level, for how it is that we sometimes feel the sufferings of another almost as if they were our own.

This metaphysical view of the underlying oneness of things is not by any means unique to Schopenhauer. He finds it expressed in the Vedas of Hinduism, in Buddhism and in the mystical expressions of most of the world's religious traditions. He would probably also have found it in the worldviews of many indigenous peoples had he been familiar with them.[43]

According to this view, individual persons *appear* to be different, just as individual islands in an ocean appear to be different and separate – but this is only the way things appear on the surface. Down at the deep ground of things, what appear on the surface to be individual beings (or islands) are discovered below simply to be upcroppings from the same underlying ground. Schopenhauer says:

> This doctrine teaches that all plurality is only apparent, that in all the individuals of this world, however infinite the number in which they exhibit themselves successively and simultaneously, there is yet manifested only one and the same truly existing essence, present and identical in all of them. Such a doctrine, of course, existed long before Kant; indeed it might be said to have existed from time immemorial … it is the main and fundamental teaching of the oldest book in the world, the sacred Vedas.[44]

The egoist, though, i.e., the person acting only in his own self-interest, fails to recognise that we are all part of the same underlying essence, and instead believes that he is an entirely separate being, and that his interests

---

[43] One example of this indigenous understanding of the world, expressed in terms that Schopenhauer might well have appreciated, derives from the world-view of the Onondaga people in upper New York state and south-eastern Canada, one of the Native American tribes in the Haudenosaunee (Iroquois) confederacy of nations.

Oren Lyons is one of today's elders, a Wisdom Keeper, in the Onondaga nation. In his youth he was the first Onondagan to enter college. His friend Huston Smith tells the story: 'When [Oren] returned to his reservation for his first vacation, his uncle proposed a fishing trip on a lake. Once he had his nephew in the middle of the lake where he wanted him, he began to interrogate him. "Well, Oren," he said, "you've been to college; you must be pretty smart now from all they've been teaching you. Let me ask you a question. Who are you?" Taken aback by the question, Oren fumbled for an answer. "What do you mean, who am I? Why, I'm your nephew, of course." His uncle rejected his answer and repeated his question. Successively, the nephew ventured that he was Oren Lyons, an Onondagan, a human being, a man, a young man, all to no avail. When his uncle had reduced him to silence and he asked to be informed as to who he was, his uncle said, "Do you see that bluff over there? Oren, you are the bluff. And that giant pine on the other shore? Oren, you are that pine. And this water that supports our boat? You are this water"' (recounted in H Smith, *The World's Religions*, Revised and updated [Kindle Edition] (HarperCollins 2009), Kindle location 7771–88).

[44] Schopenhauer, *On the Basis of Morality* (1995) (n 9) 207.

are entirely separate from the interests of other selves who appear separated from him in time and space. The egoist sees only the phenomenal appearances of things shaped by temporality, spatiality and plurality. Other persons appear to him as completely separate beings with wants and needs entirely separate from his. This illusion, according to Schopenhauer, is what accounts for egoism.

On the other hand, the person acting from compassion,

> makes less of a distinction than do the rest between himself and others … In fact, in magnanimous deeds it [this distinction] appears to be abolished, since here the weal of another is protected and supported at the expense of the good [person], and thus another's ego is treated as equal with his own.[45]

So the egoist and the person of compassion are acting on the basis of two entirely different perceptions of reality, one considering separateness and individuation to be real, the other considering them illusory.[46] But which of the two is correct?[47]

For Schopenhauer, the answer is that plurality and separateness belong only to the world of appearance and are not part of the fundamental oneness of things.

> [I]f plurality and separateness belong only to the *phenomenon*, and if it is one and the same essence that manifests itself in all living things, then that conception that abolishes the difference between ego and non-ego [i.e., compassion] is not erroneous; but on the contrary, the opposite conception must be … Accordingly, [compassion] would be the metaphysical basis of ethics and consist in *one* individual's again recognizing in *another* his own self, his own true inner nature. Thus doing right and doing good, would in the end harmonize perfectly with the profoundest teaching.[48]

The actions of egoism, on the other hand, express the illusory belief that:

---

[45] Ibid 204–205.

[46] 'That the world has only a physical and not a moral significance is a fundamental error, one that is the greatest and most pernicious, the real perversity of mind'; from Schopenhauer's essay *On Ethics*, quoted in G Mannion, *Schopenhauer, Religion and Morality: The Humble Path to Ethics* (Aldershot, Ashgate 2003) 191.

[47] 'The question now is whether this latter conception of the relation between one's own ego and another's, which is the basis of the actions of a good character, is mistaken and due to a delusion, or whether such is rather the case with the opposite conception on which egoism and malice are based' (Schopenhauer, *On the Basis of Morality* (1995) (n 9) 205, Schopenhauer's emphasis).

[48] Ibid 209, Schopenhauer's emphasis.

> 'Individuation is real; the *principium individuationis* and the diversity and variety of individuals based on this are the order of things-in-themselves. Each individual is a being radically different from all others. In my own self alone I have my true being; on the other hand, everything else is non-ego and foreign to me'. This is the knowledge to whose truth flesh and bone bear witness; it lies at the root of all egoism and is really expressed in every loveless, unjust, and malicious action.[49]

So the egoist's actions, says Schopenhauer, express the belief that individuation is real and substantial, while the compassionate person's actions express the recognition that individuals are not as separate as they appear.[50] The actions of the egoist spring from illusion, while the actions of the just and compassionate person arise from recognising the true nature of things.[51]

## 9. OUTRAGE

But how does all this account for the experience of moral outrage? Schopenhauer refers to two grotesque events that he thinks would outrage anyone because they are so deeply cruel.

> … for example [says Schopenhauer], the case recently reported in the papers, of a mother who murdered her five-year-old son by pouring boiling oil down his throat and her younger child by burying it alive; or the case, just reported from Algiers, where, after a casual dispute and fight between a Spaniard and an [Algerian], the latter, as the stronger, tore away the whole of the lower jawbone of the former, and carried it off as a trophy, leaving the other man still alive; when we hear of such things, we are seized with horror and exclaim: 'How is it possible to do such a thing?'

---

[49] Ibid 210.

[50] 'Schopenhauer does tend to regard doing the right thing as coincidentally engaging in what we might call "applied metaphysics"'. MA Fox, '"Boundless Compassion": The Contemporary Relevance of Schopenhauer's Ethics' (2006) *Eur Leg* 378–79. Schopenhauer writes, 'To be just, noble, and benevolent is nothing but to translate my metaphysics into actions' (quoted in Fox 372).

[51] 'Good characters make less of a distinction than do the rest between themselves and others, treating others as an "I once more", whereas bad characters treat others as an absolute "non-I". From a strictly empirical perspective, he continues, it appears as if the bad person's attitudes and behavior are correct, because each person is spatially distinct, and so it seems as if others are absolute "non-I's". But the empirical standpoint itself is not warranted, Schopenhauer claimed …', DE Cartwright, *Schopenhauer: A Biography* (n 12) 493.

Schopenhauer then examines what we mean when we ask this question:

> What is the meaning of this question? Is it: How is it possible to have so little fear for the punishments of the future life? Hardly. Or: How is it possible to act according to a maxim that is so absolutely unfitted to become a general law for all rational beings? Certainly not. Or: How is it possible so utterly to neglect one's own perfection and that of another? Again, certainly not. The sense of that question is certainly only this: How is it possible to be so utterly bereft of compassion? Thus it is the greatest lack of compassion that stamps a deed with the deepest moral depravity and atrocity. Consequently, compassion is the real moral incentive.[52]

The experience of outrage is thus an immediate and visceral reaction to the absence of compassion, i.e., to the absence of something we intuitively recognise as fundamental to our nature as living beings. I think Schopenhauer is correct here. When we witness or learn of acts that we find appalling, that leave us aghast and make us ask 'How could that be?', the sense of that question is 'How could anyone so utterly disregard the suffering of others and be so wilfully inattentive to the horrific impacts of their actions? Do they not see what they are doing? Have they no heart?'

The conclusion here is that it is a failure of compassion, and a lack of fellow feeling that is at odds with our nature and with our understanding of the deep nature of things, that accounts for the feeling of outrage.[53]

One additional note before we leave Schopenhauer. With regard to the question 'How is it possible to be so bereft of compassion?' one answer could be this. 'Here's how: Be a corporation'. Corporations may be 'persons' in law, but they are certainly not persons in actuality. And as Anna Grear[54] has so eloquently shown us, they do not share the rich vulnerabilities of, or sensitivities to, embodied human persons. Corporations are not, in Schopenhauer's terms, motivated by malice in the sense that they *want* to cause suffering and injury. Rather, given their mandate to maximise profits for shareholders, they are simply unambiguously and unashamedly egoistic actors in that their sole motivation is to benefit the

---

[52] Schopenhauer, *On the Basis of Morality* (1995) (n 9) 169–70.

[53] 'I think that you could also find another basis in Schopenhauer for outrage, one that lies just in the reverse of responses to compassion behavior, that is wrongful actions are morally reprehensible (one of the few times Payne went awry was rendering "*moralisch verwerflich*" as "morally bad"), eliciting strong feelings of disapprobation', personal email from David Cartwright, 6 July 2012.

[54] A Grear, *Redirecting Human Rights: Facing the Challenge of Corporate Legal Humanity* (Palgrave Macmillan 2010).

corporation and its owners. It just sometimes happens with egoism that other people's lives get in the way of egoistic purposes, and the impacted people and communities get trampled. And if those individuals and communities can get trampled on efficiently, and with little resistance, that suits the corporation's purposes even better. Outrage at their disregard for the sufferings of those who are in their way is, as in Schopenhauer's example, outrage at their being 'utterly bereft of compassion'.[55]

## 10. HUMAN RIGHTS

If Schopenhauer's ethic of compassion explains the outrage we feel at unnecessary suffering, our widely accepted international human rights tradition provides another level of validation for that sense of outrage and the felt necessity of acting on it. If compassion says: 'That suffering is horrific and unconscionable', human rights norms say: 'Yes, and it is also wrong and unjust. You are right to feel that outrage; others would too. Human suffering should not be so disregarded, and you are right to fight against it'. Human rights norms, in other words confirm the experience of compassion. They say: 'It's not just you and it's not just because you are over-sentimental. Your intuitions are correct. This is genuinely immoral and should not be accepted'. Feelings, after all, may be seen as inner, personal and private and could perhaps be interpreted as relevant only for

---

[55] This chapter is about *human* rights, not animal rights, but it is important to recognise that Schopenhauer's philosophy of compassion applies equally as well to non-human animals. 'The moral incentive that I have presented is further proved to be genuine in that it also takes under its protection *animals*, which are cared for in other European systems of morals in such an unjustifiably bad way. That animals are alleged to be without rights, the delusion that our actions toward them are without moral significance, or, as is said in the language of that morals, that there are no duties toward animals, is frankly a revolting crudity and barbarism of the Occident' (Schopenhauer, *On the Basis of Morality* (2010) (n 9) 239).

Schopenhauer's explanation for this compassion towards other animals is that 'the essential and primary thing is the same in the animal and in the human, and that what differentiates the two lies not in the primary …, in the inner essence, which in the one as well as in the other is *the will* of the individual. But what differentiates the two lies only in that which is secondary, in the intellect, in the degree of cognitive power [which results from a mere] somatic difference of a particular part, the brain' (ibid 241, Schopenhauer's emphasis). Schopenhauer concludes that 'one may confidently assert that whoever is cruel to animals could not be a good person' (ibid 242).

me. Human rights norms, on the other hand, are public, not private, and are visible, widely agreed upon and capable of being recognised and discussed in the public square.

Human rights standards, precisely because of their public nature and because of having been so widely endorsed, provide both validation for the more private world of moral intuitions, and a formal language for engaging with others in the public square about environmental wrongs that negatively impact human persons. If compassion is primarily affect (and thus more subjective), human rights norms are primarily discursive, rational and public. Compassion feels horrified and says: 'This is unconscionable'. The human rights tradition sees violation and says: 'Yes, this is morally wrong, unjust, and perhaps legally actionable'. If compassion stirs and moves the individual, human rights norms voice the judgement of the larger community, publicly confirming what the individual has felt.

The modern human rights movement, begun at the end of World War II, captured and gave voice to the moral outrage at the horrific events that had taken place a few years earlier in a modern, western-world state (Germany) with a government that had been democratically elected by an educated populace. Those gruesome events had been implemented according to legally enacted ordinances and were regulated and overseen by administratively legitimate government ministries. Those horrors were, in other words, committed in full compliance with the rule of law (as are many environmental assaults today).

We know there were some both within and outside of central Europe who felt personal outrage at these events and acted on it, but for many their outrage had to remain private and personal. The only supra-legal standards of conduct available at the time, i.e., standards that could claim to supersede laws of the state and could possibly have justified the outrage, included – but only for believers – the teachings and principles of conduct in certain religious traditions. Some, we know, did choose to act on those higher standards. But there did not yet exist any broadly accepted, formally endorsed, human rights standards or instruments, not tied to a particular religious tradition, that could claim to take precedence over national laws. There were not yet available any widely accepted public standards or fora to validate a person's inner sense of outrage. We can only guess whether anything would have been different if there had been.

The moral values expressed in the UDHR[56] and in subsequent treaties and declarations articulated a set of fundamental moral norms – more fundamental even than the laws of nation states – that could serve as 'a common standard of achievement for all peoples' (UDHR Preamble). To the extent that these declarations and treaties have been successful, they serve as external, public validation of the moral outrage felt when confronting horrific human wrongs, even when those wrongs are committed under cover of law.

## 11. HUMAN RIGHTS IMPACT ASSESSMENTS

The private inner sense of outrage at moral wrongs is confirmed, if this chapter's thesis is correct, by the existence of widely accepted and publicly endorsed human rights standards. Those standards, however, are relatively broad and generic and there may be some question as to whether a specific human rights norm would even be applicable to a given environmental situation. Does the right to security of person, for example, apply to families in rural areas who are impacted by commercial aerial pesticide sprays? Does the right of women and children to special consideration apply to families living near a hydraulic fracking operation? If a detailed Human Rights Impact Assessment (HRIA) could be prepared for a specific environmental situation that had clear adverse human impacts, that human rights assessment would provide an even more specified confirmation of that outrage.

To date Human Rights Impact Assessments can be prepared through at least three different sources: (a) Environment and Human Rights Advisory (EHRA); (b) NomoGaia, 'a nonprofit research and policy organization dedicated to clarifying the corporate role in human rights protection and facilitating corporate responsibility';[57] and (c) an HRIA Toolkit, designed to address pesticide issues, recently developed by a

---

[56] In the world of ethics, something genuinely unique came into the world with the signing and adoption of the UDHR. Never before in human history had a document about moral values been conceived, written, and endorsed by representatives of virtually every nation on earth. René Cassin, one of the drafters of the UDHR (who, for his work, was awarded the Nobel Peace Prize in 1968), is quoted as saying that, with the UDHR, 'something new … entered the world'. It was, he said, 'the first document about moral value adopted by an assembly of the human community', J Morsink, *The Universal Declaration of Human Rights: Origins, Drafting and Intent* (Philadelphia, University of Pennsylvania Press 1999) 33.

[57] http://www.nomogaia.org/.

graduate student in the Human Rights Education Program at the University of San Francisco. A few words about each.

(a) Environment and Human Rights Advisory (EHRA),[58] the organisation with which I am most familiar, has prepared (gratis) a number of HRIAs over the past few years for specific environmental situations such as: (1) a proposed aerial pheromone/pesticide spray over several hundred square miles of California's central coast and San Francisco Bay area that would have directly impacted over three million residents;[59] (2) a proposed biomass power generation plant situated in an economically disadvantaged neighbourhood of one Oregon city;[60] and (3) a proposal to open New York state to hydraulic fracturing for natural gas.[61] These HRIAs have been commissioned by environmental organisations working on the issue and have been addressed to the head of the agency or corporation with authority to make decisions about the proposed environmental action. The HRIA of hydraulic fracturing for natural gas, for example, was commissioned by Earthworks' Oil and Gas Accountability Project in Washington DC, and was addressed to the New York State Department of Environmental Conservation.

These HRIAs summarise the relevant facts and concerns in the situation, lay out a number of specific human rights norms of concern, and detail how each norm applies to that situation. One of the norms of concern in the New York fracking report,[62] for example, is 'Motherhood and childhood's right to special care'.[63] The HRIA describes what this right entails and what the specific reasons for concern are with regard to hydrofracking, and then does the same for 25 additional norms of concern. It then details a set of potential liabilities the New York State Department of Environmental Conservation could face if human rights

---

58 http://environmentandhumanrights.org. I am the founder and director.

59 http://www.lbamspray.com/00_Documents/2008/EHRA.pdf.

60 http://www.environmentandhumanrights.org/resources/EHRA%20Seneca_Eugene%20report%20090828_1.pdf.

61 http://www.environmentandhumanrights.org/resources/EHRA%20frac%20rpt%20111212-1.pdf.

62 T Kerns, 'A Human Rights Assessment of Hydraulic Fracturing for Natural Gas' (Environment and Human Rights Advisory 2011), http://environmentandhumanrights.org/resources/EHRA%20frac%20rpt%20111212-1.pdf (accessed 23 May 2012).

63 This right is articulated in UDHR Article 25, Covenant on Economic, Social and Cultural Rights, Article 12, Convention on the Rights of the Child, Article 27 and United Nations Declaration on the Rights of Indigenous Peoples, Article 22.

standards were to be abridged, and what specific measures the Department could take to reduce those liabilities.

These HRIAs are addressed to the responsible agency or corporation and are presented as purely informational, just as Environmental Impact Assessments and Health Impact Assessments are. The assessments say, in effect: 'Here are some liabilities you may be facing with respect to these specific human rights norms, and here are some specific measures you can take to reduce those liabilities'. More directly, though, these HRIAs can benefit activists who are engaging the issue, providing them with a public, verifiable set of specific human rights standards applicable to their situation and showing that the outrage they feel is appropriate, justifiable and supported by human rights norms.

(b) HRIAs developed by NomoGaia appear to serve similar purposes, and are designed to alert both the corporation and the rights-holders in the area who may be adversely impacted by the corporation's activities as to the human rights norms that apply to their situation. NomoGaia has detailed the process and results of providing HRIAs to corporations in two different situations.[64]

(c) An HRIA Toolkit that focuses specifically on pesticide-related issues, an HRIA/P, has been developed by Emily Harden, a graduate student in the Master of Arts in Human Rights Education at the University of San Francisco. Ms Harden, in collaboration with EHRA and with Pesticide Action Network North America, developed a Toolkit that includes a Guidebook with HRIA/P instructions and a formatted template for the full HRIA/P as well as a secondary template for a briefer version of the HRIA/P to be designed for distribution among the affected rights-holder community.[65] A key element of this HRIA/P process involves active engagement with the rights-holder community in developing and validating the final HRIA/P itself as well as in validating its numerical scoring of the degree and likelihood of impact for each human rights norm at issue.[66]

---

[64] K Salcito et al., 'Assessing Human Rights Impacts in Corporate Development Projects' (2013) *Environ Impact Asses* 39–50.

[65] E Harden, 'HRIA Toolkit for Pesticide Issues' (Environment and Human Rights Advisory 2014), http://www.environmentandhumanrights.org/ehra-reports-database/hriaptoolkit/ (accessed 15 August 2014).

[66] Private corporations may also choose to prepare their own HRIAs, of course, both to internally review possible human rights impacts of a given project, and to be prepared in the event of public criticism on grounds of human rights violations.

A fourth model for a Human Rights Impact Assessment that addresses an environmental situation is the 2008 climate change report prepared by the International Council on Human Rights Policy, *Climate Change and Human Rights: A Rough Guide*. Though its structure varies from those above, its purposes are similar.

Thus, just as common-sense perceptions about matters of fact in environmental situations need to be verified and supported by reference to scientific studies, so personal convictions about the moral wrongness of a situation need to be verified and validated by reference to another kind of external, public and broadly verifiable norm. HRIAs are intended to serve that purpose.

## 12. PRACTICAL IMPLICATIONS

If the central claims of this chapter have validity, if Lynn Hunt's historical account of the role of empathy in the genesis of rights discourse is accurate and if she is correct to argue that 'rights are best defended in the end by the feelings, convictions, and actions of multitudes of individuals, who demand responses that accord with their inner sense of outrage',[67] then at least three practical implications follow for environmental activism.[68]

- *Telling the story*. Simple, meaningful personal accounts of direct impacts that environmental assaults have had on individuals, families and communities should be collected, documented and widely publicised in order to help awaken the moral imagination and to evoke the compassion and outrage that will be necessary for the emergence of genuine change. Voice of Witness,[69] a non-profit organisation that focuses on the power of personal story in 'illuminating human rights crises', can serve as a source for methodologies and inspiration in this work.

---

[67] Hunt (n 5) 213.

[68] In addition to the practical measures discussed below, several others are outlined on the 'Human Rights Methods and Practice' page of my online course in Environment and Human Rights, http://environmentandhumanrightscourse.info/lecsite/practicalmethods.html.

[69] http://voiceofwitness.org.

Claims of injury or harm in these stories may need to be substantiated by reference to scientific studies and/or expert testimony. Independently commissioned Environmental Impact Assessments, Health Impact Assessments and Economic and Social Impact Assessments will be helpful in validating these claims.

- *Claiming moral authority*. Providing activists with well-researched Human Rights Impact Assessments of their specific situation can help clarify the moral dimensions and values at stake as well as identifying potential legal pressure points. As noted above, foregrounding and documenting human rights standards that apply in that particular situation can publicly validate the felt sense of injustice and legitimise the outrage experienced by those whose lives have been negatively impacted.
- *Exercising moral power*. Community-led public inquiries, such as New Zealand's 'People's Inquiry into the impacts and effects of aerial pesticide spraying over urban areas of Auckland' held in 2006,[70] and public tribunals like the Permanent People's Tribunal (2011) Session on Agrochemical Transnational Corporations held in Bangalore, India in December 2011, can be both educative and powerful. Since both formats, inquiries and tribunals, are community-initiated and led (not government-initiated), they can be structured to ensure that the voices of those who have been impacted are adequately heard.

## 12.1 People's Tribunals

A People's Tribunal is a community-led, quasi-litigation event, similar to a courtroom proceeding, in which a complainant or class of complainants who have suffered injuries due to environmental exposures tries an agency, corporation or other defendant deemed responsible for those exposures. The agency, corporation or other defendants are judged, not against state laws but against moral and legal standards expressed in international human rights instruments.

A community, for example, may try an agricultural or forestry enterprise or government agency that aerially applies pesticides in the near vicinity of their children's schools, for violation of human rights standards designed to protect children's safety and health. Or a community

[70] J Goven et al., 'Report of the March 2006 People's Inquiry into the impacts and effects of aerial spraying pesticide over urban areas of Auckland' (The People's Inquiry Inc 2007), http://peoplesinquiry.files.wordpress.com/2013/09/web-report-finalreformatted01-1.pdf (accessed 19 October 2013).

may try the regulatory agency for failure to adequately regulate, or for failure to sufficiently protect the community's health.

The best example of this to date is the Permanent People's Tribunal (2011) Session on Agrochemical Transnational Corporations held in Bangalore, India. This Tribunal, comprised of respected jurists from around the world, sat for four days in December, 2011 and considered indictments against the six largest agrochemical corporations (Monsanto, Syngenta, Bayer, Dow Chemical, DuPont and BASF) and their home states (United States, Switzerland and Germany) on charges of human rights violations. The Tribunal heard testimony from victims, survivors, witnesses and experts in various fields. When the verdict was issued several months later, the corporations and their home states were found 'responsible for gross, widespread and systematic violations of the right to health and life, economic, social and cultural rights, as well as of civil and political rights, and women and children's rights' (Permanent People's Tribunal Verdict).[71]

The primary purposes of a People's Tribunal are education about human rights standards and environmental hazards, and perhaps also the mobilisation of shame directed against indicted parties. The panel will be comprised of three to five jurists of high standing familiar with international human rights standards, and the venue should be a place of dignity and authority. The geographical location, if possible, should be near where the exposures and events in question occurred, partly for the convenience of witnesses, partly to maximise media coverage, and partly with an eye to the location's symbolic power. The War Crimes Tribunals after World War II, for example, were held at Nuremberg partly for symbolic reasons because Nuremberg was a centre of power during the Third Reich.

The primary differences between a Peoples' Tribunal and normal courts of law are:

- A Peoples' Tribunal does not have the formal legal standing, and its findings do not have the compulsory legal force of governmental or intergovernmental courts.
- The standards against which defendants in a People's Tribunal are judged are those expressed in international human rights documents, rather than those expressed in domestic laws.

[71] http://www.agricorporateaccountability.net/en/page/ppt/167 (accessed 18 March 2015).

## 12.2 People's Inquiries

Community-led People's Inquiries are similar to Tribunals in form, venue and purpose, but differ from Tribunals in that they are less like trials that result in judgments and verdicts, and more like a public fact-finding inquiry, or a Truth and Reconciliation Commission.

An Inquiry meets for several days in a public venue to hear testimony, examine evidence, deliberate and in time issue a final report that tells the community's story from the point of view of those who have testified and others they represent. Its Final Report draws conclusions from this evidence and makes recommendations.

The only example so far of such an Inquiry, aside from the Truth and Reconciliation Commissions held in some parts of Africa and South America, is the People's Inquiry held in Auckland, New Zealand in February 2006[72] (on which I served as one of the four commissioners). Its purpose was to examine the government's handling of an aerial spray programme in which helicopters and fixed wing aircraft had, for two and a half years, regularly deployed liquid pesticides over several thousand acres of urban Auckland in an attempt to eradicate a Painted Apple Moth population.

Details about the arrangements, structure, venue, stated purposes and format of that Inquiry, as well as to download its Terms of Reference, Final Report, and the collection of much of the formal testimony on which the report was based, can be found on the People's Inquiry website.[73] Part III of that Final Report includes an analysis of the spray programme from a human rights perspective.

Though bringing human rights norms to bear on environmental concerns does have certain practical advantages,[74] it is not suggested here that these methods will produce immediate results. As Martin Luther King has so eloquently reminded us, '[t]he arc of the moral universe is long but it bends toward justice'.[75]

---

[72] Ibid.

[73] http://peoplesinquiry.wordpress.com.

[74] T Kerns, 'Ten Practical Advantages of a Human Rights Approach to Environmental Advocacy' (2013) *Journal of Environmental Studies and Sciences* 416–20; JH Knox, 'Climate Ethics and Human Rights' (2014) *Journal of Human Rights and the Environment* 22–34.

[75] ML King, 'Sermon at Temple Israel of Hollywood' (American Rhetoric Online Speech Bank 1965), http://www.americanrhetoric.com/speeches/mlk templeisraelhollywood.htm (accessed 2 November 2013).

## 13. FINAL THOUGHTS

Schopenhauer scholar David Cartwright suggests that if there are differences between masculine and feminine approaches to ethics, then Schopenhauer's ethic may be closer to the feminine approach than to the masculine approach characteristic of classical philosophers like Aristotle, Kant and Mill:

> If it is true, as Carol Gilligan has argued in *In a Different Voice*, that a woman's morality differs from a man's morality in being more concrete and contextual than abstract, … more personal rather than impersonal, motivated more by care than duty, and structured more by responsibilities than rights, one could mount a strong argument that the misogynistic Schopenhauer has more of a woman's morality than a man's.[76]

Schopenhauer's philosophy is also consistent with Lynn Hunt's claim, in *Inventing Human Rights: A History*, that the empathy that paves the way for the emergence of a human rights movement 'depends on the recognition that others feel and think as we do, that our inner feelings are alike in some fundamental fashion',[77] and that this experience of empathy with others 'serves as the foundation of human rights'.[78] This chapter's thesis is also consistent with Hunt's recognition that '[h]uman rights are not just a doctrine formulated in documents; they rest on a *disposition toward other people, a set of convictions about what people are like* and how they know right and wrong in the secular world'.[79] Hunt argues that this sense of empathy arose at a certain point in history, while Schopenhauer regards empathy as a fundamental characteristic of human nature – two views which may not be as inconsistent as might appear at first glance.

This chapter's thesis is also consistent with Alan Dershowitz's argument in *Rights from Wrongs: A Secular Theory of the Origins of Rights*,[80]

---

[76] Cartwright continues: 'Although Schopenhauer argues from stereotypic grounds that justice is more a masculine virtue, philanthropy more a feminine virtue … he also believed that philanthropy or loving-kindness is the highest virtue' (Cartwright, in Schopenhauer, *On the Basis of Morality* (1995) (n 9) xxviii).

[77] Hunt (n 5) 29.

[78] Ibid 32.

[79] Ibid 27, my emphasis.

[80] 'This book is respectfully dedicated to the countless victims of terrible human wrongs – wrongs that have been the source of human rights. May these rights help to prevent the recurrence of these and other wrongs': A Dershowitz,

which sees rights as growing out of our experience of great wrongs perpetrated against humans. Dershowitz's theory, like Schopenhauer's, considers itself to be based in empirical observation rather than in *a priori* analyses of the nature of human good or of the best life.[81] Dershowitz claims that rights are 'an experiential reaction' to 'agreed-upon wrongs of the past that we want to avoid repeating'.[82] Rights, in his words, 'come from human experience, particularly experience with injustice'.[83]

One advantage of Dershowitz's theory is that it does help explain the emergence of so-called 'new' rights, such as the right to clean air and to clean and adequate drinking water. These rights, he would say, have arisen out of our experience in the past half-century with so many communities being deprived of clean air and water.

Schopenhauer's theory, like Hunt's, is also consistent with the well-recognised importance of personal narratives from the victims of environmental human rights abuses. Without such narratives it is virtually impossible for others to adequately appreciate the impacts that these abuses have had on individuals, families and communities.

---

*Rights from Wrongs: A Secular Theory of the Origins of Rights* (New York, Basic Books 2005 v–vi).

[81] 'I reject Aristotle's argument that we cannot define rights without first determining "the nature of the most desirable way of life". It is enough to have a conception – or a consensus – about the very bad society, and about the wrongs that made it so. Based on this experience with wrongs, rights can be designed to prevent (or at least slow down) the recurrence of such wrongs. … It is more realistic to try to build a theory of rights on the agreed-upon wrongs of the past that we want to avoid repeating, than to try to build a theory of rights on ideal conceptions of the perfect society about which we will never agree. Moreover, a theory of rights as an experiential reaction to wrongs is more empirical, observable, and debatable, and less dependent on unprovable faith, metaphor, and myth, than theories premised on sources external to human experience' (ibid 6, 7).

[82] Ibid 7.

[83] Ibid 8.

PART 4

# Multilevel reformulations

# 9. Reimagining ecological governance through human rights and a rediscovery of the Commons

**David Bollier and Burns H Weston**

## 1. INTRODUCTION

One of the most distressing realities of our time is the failure of governance institutions to confront well-documented ecological and social problems with vigour, honesty, and imagination. A kind of political and institutional paralysis is precluding the growth of promising alternatives.

In this chapter, we argue that the Commons paradigm, especially when guided by human rights law and policy, can help us imagine and implement a new vision of provisioning and democratic governance. We do not propose an abrupt, revolutionary change, but one that can feasibly evolve within the fragile, deteriorating edifice of existing institutions.

The Commons as a discourse and set of ethical social practices offers many attractive features in navigating a transformation in ecological governance. It offers a coherent economic and political critique of existing State/Market institutions,[1] and its history embraces many venerable legal principles that help us both to imagine new forms of law and to

---

[1] While we recognise the validity of the conceptual distinctions that customarily are made between the 'State', and 'Market,' each serving different roles and formally separate from the other, we also recognise the existence of an intimate operational alliance between these two social constructs (distinct from, and often at odds with, 'consumers' or 'consumer society') that plays a dominant role in contemporary global governance, ecological and otherwise. For convenience, therefore, we invoke the term 'State/Market' to refer to this close symbiotic relationship when, in our view, the context of its use reflects a deep, shared commitment to a political and economic agenda (neoliberal at present) and an active collaboration to advance it. We do not mean to suggest that there are not significant variations in how the State and Market interact from one nation to

develop proactive strategies for effecting change.[2] Perhaps most importantly, the Commons is supported by an actual transnational movement of commoners who are co-creating innovative new provisioning and governance systems that work.[3]

It is important to understand that the Commons is not an ideological agenda or an impractical, utopian vision. It is a useful new/old framework and vocabulary for building a new societal vision and for imagining constructive alternatives to the neoliberal economics and policies that now enclose (commodify and privatise) shared resources at virtually every opportunity, dispossessing communities and degrading our natural environment and public order.

The Commons is not just an intellectual construct, however. It is equally a constellation of self-organised projects around the world that are meeting human needs in equitable and ecologically responsible ways. They include such diverse endeavours as seed-sharing cooperatives in

---

another. But the general alliance between the two in promoting economic growth, consumer demand, and material 'progress' as overriding if not exclusive goals is, however contingent, unmistakable.

[2] This theme is extensively discussed in BH Weston and D Bollier, *Green Governance: Ecological Survival, Human Rights and the Law of the Commons* (New York, Cambridge University Press 2013), esp. chs 7–8, at 179–268. See also P Linebaugh, *The Magna Carta Manifesto: Liberty and Commons for All* (Berkeley CA, University of California Press 2008); additionally, D Bollier, *Think Like a Commoner: A Short Introduction to the Life of the Commons* (Gabriola Island BC, New Society Publishers 2014); D Bollier and S Helfrich (eds), *The Life of the Commons: A World Beyond Market and State* (Amherst MA, Levellers Press 2012), esp. Part II, 'Capitalism, Enclosure and Resistance', at 114–216; H Reid and B Taylor, *Recovering the Commons: Democracy, Place and Global Justice* (Urbana IL, University of Illinois Press 2010); M Haiven, *Crises of Imagination, Crises of Power: Capitalism, Creativity and the Commons* (London UK, Zed Books 2014); T Schroyer, *Beyond Western Economics: Remember Other Economic Cultures* (New York NY, Routledge 2009); and K Polanyi, *The Great Transformation: The Political and Economic Origins of Our Time* (Boston MA, Beacon Press 1944/1957).

[3] The vitality of this transnational activism on the commons can be seen in Bollier and Helfrich (eds) (n 2); in two major conferences on the commons in Berlin hosted by the Heinrich Böll Foundation, the International Commons Conference in 2010 (http://www.boell.de/en/node/277225) and the Economics and the Commons Conference in 2013 (http://www.boell.de/en/economics-and-commons); and in the blog postings on Bollier.org, the Peer to Peer Foundation (http://blog.p2pfoundation.net) and the Commons Blog (https://commonsblog.wordpress.com), among others (all accessed 1 October 2014).

India,[4] water-sharing systems throughout the global South,[5] communities of open source software programmers producing shareable code,[6] localities that use alternative currencies to invigorate their economies,[7] the 'right to the city' movement in Europe that defends public space from overdevelopment,[8] and local food initiatives such as Community Supported Agriculture, Slow Food, and permaculture.[9] As a system of provisioning and governance, the Commons lets people make their own rules for managing the resources on which they depend. It gives them a significant degree of sovereignty and control in the spheres of everyday life that matter to them.[10] It helps them reconnect to nature and to each other, set limits on resource exploitation, and internalise the 'negative externalities' so often associated with Market behaviour.[11]

---

[4] The group founded by Indian activist Vandana Shiva, Navdanya (http://www.navdanya.org) is quite active in promoting seed-sharing. See also D Bollier, 'The Seed Sharing Solution' [blog post], Bollier.org, 19 Jan 2011, at http://bollier.org/seed-sharing-solution (each accessed 1 October 2014).

[5] See e.g., A Davidson-Harden, 'Alternatives for Local Control of Water Commons: Stories of Rising to the Challenge' [report] (Council of Canadians, OurWaterCommons.org, et al.), http://www.ritimo.org/IMG/pdf/Local-Control-and-Management-of-Our-Water-Commons-1.pdf (accessed 1 October 2014).

[6] There is a large and diverse literature on this topic, but these volumes are excellent: EG Coleman, *Coding Freedom: The Ethics and Aesthetics of Hacking* (Princeton NJ, Princeton University Press 2013); CM Kelty, *Two Bits: The Cultural Significance of Free Software* (Durham NC, Duke University Press 2008); G Moody, *Rebel Code: Inside Linux and the Open Source Revolution* (Cambridge MA, Perseus Publishing 2001); S Weber, *The Success of Open Source* (Cambridge MA, Harvard University Press 2004); and S Williams, *Free as in Freedom: Richard Stallman's Crusade for Free Software* (Sebastopol CA, O'Reilly 2002).

[7] G Hallsmith and B Lietaer, *Creating Wealth: Growing Local Economies with Local Currencies* (Gabriola Island BC, New Society Publishers 2011).

[8] See Wikipedia, 'Right to the City', https://en.wikipedia.org/wiki/Right_to_the_city (accessed 1 October 2014).

[9] For more on these movements, see Local Harvest, http://www.localharvest.org/csa; Slow Food movement, at Wikipedia, https://en.wikipedia.org/wiki/Slow_Food; and the Worldwide Permaculture Network, at http://www.permacultureglobal.com (all accessed 1 October 2014).

[10] These are primary themes developed by Elinor Ostrom in *Governing the Commons: The Evolution of Institutions for Collective Action* (New York NY, Cambridge University Press 1990), and by the international network of commons scholars associated with the International Association for the Study of the Commons (http://www.iasc-commons.org (accessed 1 October 2014).

[11] Bollier develops these themes in *Think Like a Commoner* (n 2); and with Weston in *Green Governance* (n 2), especially Part II, at 123–268.

While most commons in the classic sense are small in scale, it is entirely reasonable to extend the principles of the Commons to larger platforms such as the Internet and the State. The Internet enables commons-based collaborations on a scale involving tens of thousands of people and more. And, given appropriate guidelines, the State can credibly act as a 'State Trustee Commons' to govern larger-scale natural systems and infrastructures as warranted. Seen through this lens, there is a vast, under-theorised realm of large-scale commons that merit our attention, particularly if we are serious about addressing climate change.

To be sure, the Commons paradigm has limitations. Instigating new collective management systems, for example, can be difficult; it can be challenging to stabilise the rules, practices and culture of a given commons. What is more, it is not unusual for the State and Market to be hostile adversaries rather than supportive partners of the Commons, particularly since the Commons often competes with them by diverting revenues, loyalties, or legitimacy from one or both. It is therefore not surprising that the State and Market have historically colluded to enclose commons (generally to convert shared resources into privatised commodities) or sought to disrupt the social relationships and identities that constitute a commons lest a competing power centre arises.

Notwithstanding such complications, the Commons provides a useful template for diverse communities to imagine very different structures for governance. In a commons, ordinary people can deliberate with each other and have their concerns heard – a capacity that is sorely missing in contemporary political life and global markets,[12] but precisely what is needed if we are going to achieve any durable environmental or economic sustainability: the ability of people to formulate and ratify the rules that will affect their everyday lives. Because the Commons honours use-value over exchange-value and seeks to ensure that basic needs are met first, it avoids the Sisyphean task of ensuring business profitability as the prerequisite for tackling any social goal.

This, arguably, is the 'secret potency' of commons over time: they can help to disrupt the State/Market's pathological dependence on economic growth and begin to set enforceable limits on the overexploitation and abuse of Earth's natural systems ('the tragedy of the market').[13] Commons generally strive to nurture an ethic of sufficiency while meeting

---

[12] This theme is notably developed by Ivan Illich in his works, such as *Deschooling Society* (London, Marion Boyars 1971); *Tools for Conviviality* (London, Marion Boyars 1973); and *Medical Nemesis* (London, Marion Boyars 1976), among others.

[13] These themes are explored at length in Bollier and Helfrich (eds) (n 2).

basic needs, and so offer a pathway for escaping the doomed quest for ceaseless economic growth.

## 2. THE COMMONS PARADIGM IN BRIEF

The Commons has been a familiar if not pervasive mode of provisioning and governance since the dawn of human existence. Reciprocal altruism and collective action most certainly contributed to the development of prehistoric agriculture, and indigenous peoples around the world have ingeniously blended their cultural practices with ecosystem imperatives. This pattern continues in analogous form on the Internet where social collaboration with scant regard for private property rights or profit is virtually a default economic and social norm.[14] Indeed, a growing body of scientific evidence suggests that social trust and cooperation may be an evolutionary reality hard-wired into the human species.[15]

It is helpful to see that the Commons is really a *social* paradigm. In its very essence, the concept challenges some basic premises of conventional economic theory developed in the eighteenth and nineteenth centuries. The Commons refutes the idea that human beings are essentially selfish, rational, utility-maximising individuals. It argues that human beings have many more complex propensities – for cooperation, creative initiative, sacrifice for collective interests, idealism, etc. – that are consequential to economic activity and life. While people are surely self-interested and competitive in many aspects of their lives, they also exhibit deep concern for fairness, participation, social connection, and peer approval.

All of these human traits lie at the heart of the Commons. Yet most economists view such traits as incidental to Market transactions, the esteemed 'main act' of wealth creation, because they tend not to conform to the basic logic of *homo economicus* and Market economics.[16] The

---

[14] DJ Kennet and B Winterhalder (eds), *Behavioral Ecology and the Transition to Agriculture* (Berkeley CA, University of California Press 2006).

[15] S Bowles and H Gintis, *The Cooperative Species: Human Reciprocity and Its Evolution* (Princeton NJ: Princeton University Press 2011); MA Nowak, *Super Cooperators: Altruism, Evolution and Why We Need Each Other to Succeed* (New York, Free Press 2011).

[16] There is a large literature refuting the idea of *Homo economicus*, particularly in behavioral economics and complexity theory economics but also in realms outside of economics such as anthropology and social psychology. See e.g., D Kahneman, *Thinking, Fast and Slow* (Farrrar Straus Giroux, 2011); R Nelson, *Economics as Religion: From Samuelson to Chicago and Beyond* (Pennsylvania State University, 2001); ED Beinhocker, *The Origin of Wealth:*

human propensity to cooperate and coordinate with others, without the involvement of money, legal contracts or coercion, is in fact a powerful force in human evolution and civilisation.[17]

Remarkably, this reality has been largely dismissed for more than forty years. A large segment of the educated public reflexively regards the Commons as a failed management regime. Much of the blame can be traced to a famous essay written by biologist Garrett Hardin in 1968, 'The Tragedy of the Commons', a short piece in the journal *Science* that presented a parable of a shared pasture on which no single herder has a 'rational' incentive to limit his cattle's grazing.[18] Hardin argued that when a resource is shared, as in a commons, no individual has the incentive to limit her or his consumption of the resource. This invariably results in its overuse and ruin – the so-called 'tragedy of the commons' – Hardin contended.

Alas, Hardin was not describing a commons, but, rather an open access regime or free-for-all in which everything is free for the taking without constraint. In a commons, however, there is a distinct community of users who govern the resource. Among other acts, the commoners negotiate their own rules of usage, assign responsibilities and entitlements, and set up monitoring systems and penalties to identify free riders in order to maintain the Commons. Commons scholar Lewis Hyde has puckishly

---

*Evolution, Complexity and the Radical Remaking of Economics* (Harvard Business School Press, 2006). See also R Skidelsky and E Skidelsky, *How Much is Enough? Money and the Good Life* (New York NY, Other Press 2012); U Mäki, *The Economic World View: Studies in the Ontology of Economics* (New York, Cambridge University Press 2001); and F Habermann, 'We Are Not Born Egoists', in Bollier and Helfrich (eds) (n 2), at 13–18.

[17] See e.g., EO Wilson, *The Social Conquest of Earth* (New York, WW Norton 2013); MA Nowak, *Super Cooperators: Altruism, Evolution and Why We Need Each Other to Succeed* (New York, Free Press 2011); E Sober and DS Wilson, *Unto Others* (Cambridge MA, Harvard University Press 1998); S Bowles and H Gintis, *The Cooperative Species: Human Reciprocity and Its Evolution* (Princeton NJ, Princeton University Press 2011); and H Gintis and S Bowles et al., *Moral Sentiments and Material Interests: The Foundations of Cooperation in Economic Life* (Cambridge MA, MIT Press 2005). See also books on the culture and politics of cooperation: N Henrich and J Henrich, *Why Humans Cooperate: A Cultural and Evolutionary Explanation* (Oxford, Oxford University Press 2007); R Sennett, *Together: The Rituals, Pleasures and Politics of Cooperation* (New Haven CT, Yale University Press 2012).

[18] G Hardin, 'The Tragedy of the Commons' (Dec 1968) 162/3859 *Science* 1243.

called Hardin's 'tragedy' thesis 'The Tragedy of Unmanaged, Laissez-Faire, Commons-Pool Resources with Easy Access for Non-Communicating, Self-Interested Individuals'.[19]

Nonetheless, economists and conservative ideologues saw the parable as a useful way to affirm the virtues of private property and limited government. With their active support, Hardin's misrepresentation of the Commons became a familiar catchphrase, with the 'tragedy' syndrome going on to become a much-taught concept in undergraduate education.[20] In the meantime, contemporary economic theory has ignored actual working commons. Two leading American introductory economics textbooks, for example, do not explore the natural resource commons upon which an estimated two billion people (almost one-third of the world's population) rely to meet their everyday needs.[21] Subsistence provisioning for household use is not as captivating to economists as production for Market exchange and capital accumulation.

Professor Elinor Ostrom, a political scientist at Indiana University, helped rescue the Commons from the memory hole to which mainstream economics had consigned it. Over the course of four decades, Ostrom's extensive empirical fieldwork documented the capacity of communities to manage natural resources sustainably. The central question that Ostrom and her colleagues have tried to answer was:

> how a group of principals who are in an interdependent situation can organize and govern themselves to obtain continuing joint benefits when all face temptations to free-ride, shirk or otherwise act opportunistically. Parallel questions have to do with the combinations of variables that will (1) increase the initial likelihood of self-organization, (2) enhance the capabilities of individuals to continue self-organized efforts over time, or (3) exceed the capacity of self-organization to solve CPR [Common-Pool Resource] problems without external assistance of some form.[22]

---

[19] L Hyde, *Common as Air: Revolution, Art and Ownership* (New York, Farrar, Straus and Giroux 2010) 44.

[20] GN Appell, 'Hardin's Myth of the Commons: The Tragedy of Conceptual Confusions', Working Paper 8. ME Phillips, Social Transformation and Adaptation Research Institute, http://dlc.dlib.indiana.edu/dlc/handle/10535/4532 (accessed 1 October 2014).

[21] PA Samuelson and WD Nordhaus, *Economics* (17th edn, New York, McGraw-Hill 2001); JE Stiglitz and CE Walsh, *Economics* (3rd edn, New York, WW Norton 2002). See also 'Securing the Commons: Securing Property, Securing Livelihoods', *International Land Alliance*, http://www.landcoalition.org/global-initiatives/securing-commons (accessed 1 October 2014).

[22] E Ostrom, *Governing the Commons: The Evolution of Institutions for Collective Action* (New York, Cambridge University Press 1990) 42.

Ostrom's landmark 1990 book *Governing the Commons* identified some key 'design principles' for successful commons, while her other books explored ways to diversify and nest governance ('polyarchy') in order to empower bottom-up initiative and decision-making.[23] For this work, Ostrom shared the 2009 Nobel Prize in Economics, the first woman to be so honoured. Coming on the immediate heels of the 2008 financial crisis, the Nobel Prize committee apparently wished to showcase a scholar whose work demonstrated how ongoing social *relationships* play as significant a role in economics as impersonal Market *transactions.*

Ostrom (who died in 2012) and her legion of colleagues produced a vast research literature over the past two decades. Most of the scholars in this large international network focused on small-scale natural resources such as forests, fisheries, grazing lands, groundwater, and wild game, usually located in rural regions of marginalised countries. Anthropologists, sociologists, political scientists, and economists and other scholars associated with the International Association for the Study of the Commons have studied pastoralists in semi-arid regions of Africa, lobstermen in the coast coves of Maine, rubber tappers in the Amazon, and fishers in the Philippines, among hundreds of other commoners.[24]

## 3. VARIETIES OF COMMONS

While there is no definitive taxonomy or inventory of commons, it can be said that a commons arises whenever a given community decides that it wishes to manage a resource collectively, with an emphasis on fair access, use, and long-term sustainability. It is tempting to categorise commons by the type of resource involved. But this can be misleading. A 'knowledge commons' on the Internet, for example, is not simply about intangible resources such as software code or digital files; they also require physical resources such as computers, electricity and food for the computer users. By the same token, 'natural resource commons' are not just about timber or fish or corn, because these resources, like all commons, can be managed only through social relationships and shared

---

[23] E Ostrom, *Understanding Institutional Diversity* (Princeton NJ, Princeton University Press 2005); J Burber, E Ostrom et al., *Protecting the Commons: A Framework for Resource Management in the Americas* (Washington DC, Island Press 2001).

[24] See e.g., the holdings of the Digital Library of the Commons, http://dlc.dlib.indiana.edu/dlc (accessed 1 October 2014).

knowledge. '*All commons* are social, and *all commons* are knowledge commons', as commons scholar and activist Silke Helfrich puts it.[25]

Truly, a commons must be conceived as an organic blending of the resource, the distinct community that manages it, and the special rules, values, social practices and other 'soft factors' that are needed to make the system function over time. However, the fact that any given commons has so many 'floating variables' makes it difficult to propound a standard, universal typology of commons. The boundaries between commons and their contextual circumstances are blurry, creating a serious methodological challenge in identifying which dynamics of commons are defining and which are incidental.

Ostrom sought to overcome this problem by developing what she called Institutional Analysis and Development (IAD), a meta-theoretical research framework for cross-disciplinary assessment of the variables at work in a commons.[26] We have our doubts that IAD can systemically organise a vast number of variables onto the same 'ontological plane' because, seen from the 'inside', as a subjective participant in a commons that has its own complicated history, decision-making and rules, personalities, traditions, and so forth, a universal typology or taxonomy of commons seems too crudely abstract to capture the highly particularistic, lived realities of commoning. Nonetheless, it is valuable to have a rough mental map of the sprawling universe of commons. So here are a few notable ecological commons that suggest the types of functional, sustainable social systems that currently manage natural resources outside of State and Market bureaucracies:

- **The Potato Park (Peru)** is a *sui generis* regime under international law that authorises a number of Peruvian indigenous peoples to manage more than 900 genetically diverse potatoes as a 'biocultural heritage' landscape. The Potato Park is a rare instance of State law recognising collaborative stewardship of crops that would

[25] S Helfrich, 'Economics and the Commons?! Towards a Commons-Creating Peer Economy' (remarks at Economics and the Commons Conference, Berlin, Germany, 22 May 2013), http://commonsandeconomics.org/2013/06/09/silke-helfrichs-opening-keynote-towards-a-commons-creating-peer-economy. Video version: http://www.youtube.com/watch?v=LLAK_M-ayfo (each accessed 1 October 2014).

[26] E Ostrom, 'The Institutional Analysis and Development Approach', in ET Loehman and DM Kilgour (eds), *Designing Institutions for Environmental and Resource Management* (Northampton MA, Edward Elgar 1998) 68–90.

otherwise be vulnerable to enclosure through patenting of genetic knowledge, aka 'bio-piracy').[27]

- ***Acequias*** are 'bio-cultural' institutions of Hispanic-Americans in New Mexico that sustainably manage irrigation water in a very arid region. In contrast to land practices in adjacent areas, *acequias* contribute to soil and water conservation, aquifer recharge, preservation of wildlife and plant habitats, and energy conservation. The success of *acequias* stems from their commons-based system of decision-making, participation, and enforcement. All community members share responsibilities for stewardship of the water (e.g., annual cleaning of water ditches) while sharing use rights to water, a system that has worked sustainably even in periods of drought.[28]
- **The System of Rice Intensification** (SRI) is an agro-ecological system for improving the productivity of irrigated rice by changing the mix of plants, soil, water, and nutrients. SRI operates as a self-organised online network of farmers in several dozen countries. These collaborations in cyberspace have helped farmers boost yields by 20 to 100 per cent, reduce the seed required by 90 per cent, and reduce water usage by up to 50 per cent. The project is notable for blending the use of online platforms with physical resource management – a trend exemplified by other 'eco-digital commons'.[29]
- **The stakeholder trust**, inspired by the Alaska Permanent Fund, is a species of large-scale commons that distributes revenues from a shared asset (such as oil on Alaskan land) to every household in the State (roughly $1,000 to $2,000 each year). For alienable resources, this governance/management model is easily replicable. For example, stakeholder trusts have been proposed for the atmosphere ('Earth Atmospheric Trust'), oceans, and the human genome. Ecological economists in Vermont have proposed a stakeholder trust for natural resources in their State. Stakeholder trusts are a

---

[27] A Argumedo, 'The Potato Park, Peru: Conserving Agrobiodiversity in an Andean Indigenous Biocultural Heritage Area', in T Amend et al. (eds), *Protected Landscapes and Agrobiodiversity Values* (Gland, Switzerland, International Union for the Conservation of Nature 2008) 45–58.

[28] See e.g., SG Crawford, *Mayordomo: Chronicle of an Acequia in Northern New Mexico* (Albuquerque, University of New Mexico Press 1993); and S Rodriguez, *Acequia: Water Sharing, Sanctity and Place* (Santa Fe NM, School of American Research Press 2007).

[29] System of Rice Intensification, http://sri.ciifad.cornell.edu (accessed 1 October 2014).

particularly useful way to recognise our collective ownership of certain natural resources and to generate non-wage income streams that can reduce inequality – a theme long championed by innovator Peter Barnes.[30]

- **Urban commons** are increasingly named and defended as such, most notably in the violent protests to prevent Gezi Park in Istanbul from being converted into a shopping mall. Urban commons include community gardens, urban land trusts, collaborative consumption, and 'homegrown parks' (Toronto). There are many new initiatives to treat 'the city as commons' and to develop policies for a 'shareable city'.[31]

Each of these very different types of commons recognises the power of local and distributed engagement; moral, social, and ecological considerations; and the creativity and legitimacy that results from self-organised, bottom-up rule-making.

## 4. THE TRANSNATIONAL COMMONS MOVEMENT

Independent of the academic study of the Commons, a global movement of commoners began to emerge in the late 1990s and early 2000s, much of it fuelled by the World Wide Web and free/open source software. These self-identified commoners – at local, national, and transnational levels

---

[30] P Barnes, *With Liberty and Dividends for All: How to Save Our Middle Class When Jobs Don't Pay Enough* (2014); P Barnes et al., 'Creating an Earth Atmospheric Trust: A System to Control Climate Change and Reduce Poverty' (March 2008) 319/5864 *Science* 724; PH Sand, 'Public Trusteeship for the Oceans', in TM Ndiaye and R Wolfrum (eds), *Law of the Sea, Environmental Law and Settlement of Disputes* (Boston, Martinus Nijhoff 2007) 521; DE Winickoff and RN Winickoff, 'The Charitable Trust as a Model for Genomic Biobanks' (Sept 2003) 329 *New England J. Med.* 1180; and D Bollier, *The Vermont Common Assets Trust*, at Bollier.org [Blog], 10 Mar 2011, http://www.bollier.org/Vermont-common-assets-trust (accessed 1 October 2014). See also Weston and Bollier, *Green Governance* (n 2), at 245–48.

[31] Homegrown National Park, http://bollier.org/blog/homegrown-urban-parks-toronto; *Shareable and Sustainable Economies Law Center, Policies for Shareable Cities: A Sharing Economy Policy Primer for Urban Leaders* (Oakland CA, Sustainable Economies Law Center 2013), http://www.shareable.net/download-your-copy-of-policies-for-shareable-cities (each accessed 1 October 2014).

– focus on some wildly eclectic types of resources. They include subsistence agriculture,[32] stewardship of water,[33] the re-localisation of agriculture and other production,[34] open access scholarly publishing,[35] new forms of 'open source design and manufacturing' for cars, furniture, farm equipment,[36] policies for a 'shareable city',[37] scientific data sharing,[38]

---

[32] V Bennholdt-Thomsen, 'Subsistence: Perspective for a Society Based on Commons', in Bollier and Helfrich (eds) (n 2), at 82–86 and http://wealthofthe commons.org/essay/subsistence-perspective-society-based-commons (accessed 1 October 2014). See also M Mies and V Bennholdt-Thomsen, *The Subsistence Perspective: Beyond the Globalised Economy* (London, Zed Books 1999); and T Waters, *The Persistence of Subsistence Agriculture: Life Beneath the Level of the Marketplace* (Lanham MD, Lexington Books 2008).

[33] See e.g., A Snitow and D Kaufman with M Fox, *Thirst: Fighting the Corporate Theft of Our Water* (San Francisco, Jossey-Bass 2007); M Barlow and T Clarke, *Blue Gold: The Fight to Stop the Corporate Theft of the World's Water* (New York, New Press 2002); and E Royte, *Bottlemania: Big Business, Local Springs and the Battle Over America's Drinking Water* (New York, Bloomsbury 2008).

[34] See n 9.

[35] The Wikipedia entry for 'open access' has a good overview of the history and state of OA (https://en.wikipedia.org/wiki/Open_access). For a listing of open access publishers, go to Scholarly Open Access (http://scholarlyoa.com/publishers), for a listing of OA journals, go to the Directory of Open Access Journals (http://www.doaj.org). There is now an Open Access Scholarly Publishers Association (http://oaspa.org). The Open Educational Resources movement is quite diverse, but a good place to learn more is the OER Commons (http://www.oercommons.org) (all accessed 1 October 2014).

[36] See e.g., B van Abel, L Evers et al., *Open Design Now: Why Design Cannot Remain Exclusive* (Amsterdam, Bis Publishers 2011); also http://open designnow.org. The Open Source Hardware and Design Alliance (http://www.ohanda.org) is an excellent source on this topic, as is the P2P Foundation website (http://blog.p2pfoundation.net). Interesting open design and manufacturing projects include the Open Prosthetics Project (http://www.openprosthetics.org) and the Wikispeed automobile project (http://www.wikispeed.org) (all accessed 1 October 2014).

[37] Shareable.net, 'Policies for Shareable Cities: A Sharing Economy Policy Primer for Urban Leaders', 5 September 2013, at http://www.shareable.net/blog/new-report-policies-for-shareable-cities (accessed 1 October 2014).

[38] See e.g., Open Data Science, at http://opendatascience.org, and the Open Knowledge Foundation, at https://okfn.org/ (accessed 1 October 2014).

'collaborative consumption',[39] cooperatives,[40] and alternative currencies,[41] among many others.

Despite their manifest differences, commoners in these various realms tend to share deep commitments to managing and preserving shared resources in fair, inclusive, sustainable, and accountable ways. Commoners also tend to care less about theorising about the Commons than in practically improvising their stewardship of shared resources. That is, they mostly focus on *building* new models of commons-based provisioning outside of the control of the State/Market.[42]

The book *The Wealth of the Commons*, a collection of 73 essays on the Commons published in 2012, provides a sense of the scope and diversity of this activity.[43] Besides many activist initiatives and commons-based provisioning projects, notable recent developments include a major international conference in Berlin in May 2013[44] and a strategic research initiative launched by Ecuador to explore how it might build its economy on the principles of commons-based peer production.[45]

The broad appeal of commons discourse stems in no small part from the vulnerability that so many people experience at the hands of the State/Market. Enclosures of water, customary lands, genetic resources, knowledge, the Internet, and much else are all motivated by a common denominator, the neoliberal policy agenda, which sees the marketisation of everything and economic growth as self-evident imperatives. Generally

---

[39] See the Shareable.net website in general, and R Botsman and R Rogers, *What's Mine is Yours: The Rise of Collaborative Consumption* (New York, Harper Collins 2010).

[40] The literature on the co-operative movement is large and diverse, but hereinafter, for convenience, 'a human right to environment' or 'right to environment'.

[41] Interest in alternative and cryptocurrencies has been booming since the advent of Bitcoin. A good overview is JH Clippinger and D Bollier, *From Bitcoin to Burning Man and Beyond: The Quest for Identity and Autonomy in a Digital Society* (Boston MA, ID3 2014), https://idcubed.org/bitcoin-burning-man-beyond (accessed 1 October 2014).

[42] The varieties of commons-based innovation are well-represented in Bollier and Helfrich (eds) (n 2).

[43] Bollier and Helfrich (eds) (n 2).

[44] The Economics and the Commons Conference, 2013, http://www.boell.de/en/economics-and-commons (accessed 1 October 2014).

[45] The research plan produced for Ecuador's Free Libre Open Knowledge (FLOK) project can be found at http://en.wiki.floksociety.org/w/Research_Plan. A follow-through project launched by the FLOK Project director Michel Bauwens is the Commons Transition Plan, described at http://p2pfoundation.net/Commons_Transition_Plan (each accessed 1 October 2014).

speaking, however, the many victims of Market enclosures have no shared discourse for expressing the value of their commons as a source of provisioning, social connection, and identity. The Commons paradigm helps address this need for a shared discourse and increasingly serves as scaffolding for developing a critique and vision. This project naturally draws upon the striking historical parallels between contemporary enclosures and the English enclosure movement, which criminalised commoning, converted shared wealth into private property, and dispossessed people of the resources and traditions needed for subsistence.[46]

This history remains highly instructive today, especially in its legal dimensions. The civil wars that resulted in King John's signing of the Magna Carta in 1215, and the companion document the Charter of the Forest, set forth a range of legal principles recognising the right to commoning.[47] They augmented a body of commons-based law going back to Roman law, most notably the Institutes of Justinian, the first known legal recognition of common resources, in 535 CE.[48] Other bodies of commons-based law include the public trust doctrine in environmental law, which prohibits the State from selling or giving away resources that belong to the unorganised citizenry;[49] legal regimes to conserve land and other natural resources; and international treaties that govern the use of oceans, Antarctica, and space.[50]

---

[46] See e.g., Linebaugh (n 2); C Hill, *The World Turned Upside Down: Radical Ideas During the English Revolution* (New York, Penguin 1972); and Polanyi (n 2).

[47] Linebaugh (n 2).

[48] TC Sandars (transl.), *The Institutes of Justinian* (Chicago, Callaghan & Company 1876), http://www.fordham.edu/halsall/basis/535institutes.asp (accessed 1 October 2014). Follow link for Book Two, Title I.

[49] MC Wood, *Nature's Trust: Environmental Law for a New Ecological Age* (Cambridge, Cambridge University Press 2013) is an excellent history and analysis of the public trust doctrine.

[50] These legal regimes are discussed at length in BH Weston and D Bollier, *Green Governance: Ecological Survival, Human Rights and the Law of the Commons* (n 2), especially in ch. 5, 'The Commons as a Model for Ecological Governance'.

## 5. IMAGINING A NEW ARCHITECTURE OF COMMONS-BASED LAW AND POLICY

In our recent book, *Green Governance: Ecological Survival, Human Rights and the Law of the Commons*,[51] we argue that a number of powerful trends – in economics, digital technology, human rights law, and commons – are converging. This convergence is not only forcing us to reconsider many seemingly fixed premises about economic and political life, it is revealing important new principles and dynamics, particularly on open networks. It is becoming more apparent that a new paradigm of understanding is needed to see that governance, production, and consumption patterns are blending together in new ways, rendering moot many of the familiar conceptual distinctions that are made between the 'State', 'Market', and 'civil society/consumers'. In a commons, people can combine important elements of self-governance, production for use and consumption for basic needs, without a significant role by either the State or Market.

We consider this a salutary development because it is empowering in a democratic sense; it ensures a greater moral legitimacy and participation in rules that must govern our everyday practices; and it provides a credible basis for limiting economic growth while still meeting people's needs. But can our legal systems provide due recognition of and support for the Commons?

Our book seeks to develop a new architecture of commons-based law and policy – 'green governance – that we believe can help support both new and old forms of commons. Our 'commons imaginary' calls for innovative uses of State and human rights law and policy to facilitate the formation and maintenance of ecological commons. We propose that modern law should recognise the age-old paradigm of the Commons and provide affirmative recognition and support to the right to common. Our premise is that, given the demonstrable failures and limitations of the State/Market in assuring a clean and healthy environment,[52] a new regime of commons-based 'green governance' (in concert with the State where needed) could help overcome some key structural limitations of 'free markets' and conventional government.

---

[51] Supra n 2.

[52] See e.g., MC Wood, *Nature's Trust: Environmental Law for a New Ecological Age* (New York, Cambridge University Press 2013), especially 'Part I: Environmental Law: Hospice for a Dying Planet', at 1–124.

This will require that we recognise law itself as something more than the enactments of legislatures and the declarations of courts. We call for the official recognition of what we call *Vernacular Law* – the 'unofficial' social norms, procedures, and customary institutions that peer communities devise to manage their own resources. 'Vernacular', as the prominent Austrian social critic Ivan Illich pointed out, 'implies "rootedness" and "abode"' and derives from the Latin word *Vernaculum* that described 'sustenance derived from reciprocity patterns embedded in every aspect of life, as distinguished from sustenance that comes from exchange or from vertical distribution'.[53] Vernacular Law matters because commons governance depends on informal, socially negotiated rules that may not even be written down. It constitutes a form of 'cultural ballast' that gives a commons stability and self-confidence even in the absence of formal law.

Some may complain that commons based on Vernacular Law are not necessarily democratic in the sense understood by modern liberal polities. There indeed may be social inequalities and hierarchies in various commons which traditional liberals might consider alien or unacceptable. On the other hand, as Ostrom's design principles for commons confirm, successful commons tend to have modes of participation, deliberation, transparency, responsibility, and effectiveness that generally surpass anything provided by the bureaucratic State or even representative democracies in practice.

In any case, one attractive option is for the State to act as a partner with the Commons, setting the basic democratic parameters within which Vernacular Law must operate, without directly controlling how a given commons is organised and managed. The State already delegates considerable authority to corporations through corporate charters, ostensibly to serve public purposes. Why should it not use this same authority to charter commons for green governance? Such a policy approach would help to mobilise decentralised participation on the ground and provide a meaningful role for local knowledge and bottom-up innovation. Moving in this direction – to recognise and support commons as a matter of State law and policy – would obviously require innovative legal and policy structures and procedures.[54]

---

[53] I Illich, *Shadow Work* (London, Marion Boyars 1981) 57.

[54] For initial discussion and detail, see Weston and Bollier (n 2) at 191–204 ('Macro-Principles and Policies to Guide the State/Market in Supporting the Commons Sector') and ch. 8 ('Catalytic Strategies for Achieving Green Governance').

In addition to the potential efficacy of commons-based ecological governance, we believe there are significant moral and political arguments for such a paradigm shift. The most significant factor may be how green governance could provide a solid foundation for actualising human rights in more robust, stable ways than the existing State/Market order allows.

## 6. THE IMPORTANCE AND POTENTIAL OF HUMAN RIGHTS LAW AND POLICY

Human rights law and policy are indispensable to commons-based ecological governance. Clean air, water, food and other resources and ecosystems are essential to life, and respect for life is, at bottom, the most basic of all human rights. Even if climate change, nuclear proliferation, massive global poverty, and other environmental hazards were not pressing hard upon us, the need for ecological governance firmly anchored in human rights law and policy would therefore be manifest.

There are other, more recognisably pragmatic reasons why human rights law and policy are indispensable to commons-based ecological governance. At present, human rights provide the only means internationally by which individuals and civil society can gain legal standing to affect definitive international environmental law-making and law-enforcing outcomes directly and peacefully. Additionally, on the national plane, environmental demands anchored in human rights can facilitate grassroots action in defence of ecosystems and victims of environmental degradation. This is especially so in countries that provision environmental rights in their constitutions, codes, and statutes (a current rising trend).[55]

Arguably most compelling, however, are five strategic virtues that, alone or more often in combination, give human rights law and policy unparalleled potential for doing individual and societal good.

First, human rights are widely understood to constitute maximum claims on society, fundamental public order values or goods at the apex of public policy, juridically more elevated than commonplace standards, laws, or other policy choices that are subject to everyday revision or rescission for lack of such ordination. A proximate analogy is the distinction between a contractual or statutory claim and a constitutional

---

[55] See e.g., DR Boyd, *The Environmental Rights Revolution: A Global Study of Constitutions, Human Rights, and the Environment* (Vancouver, University of British Columbia Press 2012).

one. As the late legal philosopher Ronald Dworkin once famously put it, when a claimed value or good is categorised as a right, it 'trumps' most if not all other claimed values or goods[56] – in no small measure because scores of human rights derive from the world's many religions, as is particularly apparent when it comes to the natural environment and environmental rights. Thus, when human abuse of a natural resource or ecosystem is designated the wronging of a right – as a violation of a human right to a clean and healthy environment, for example – the moral, political, and legal status of the claimed environmental interest, when balanced against competing policy objectives, is significantly enhanced. Human rights grant a moral, political, and legal legitimacy to the interest that it might not otherwise have.

Second, human rights signify 'a public order of human dignity … in which values are shaped and shared more by persuasion than by coercion, and which seeks to promote the greatest production and widest possible sharing, without discriminations irrelevant of merit, of all values among all human beings'.[57] A rights-based approach to ecological governance thus signals more than environmental protection *per se*; it signals also principles of freedom, nondiscrimination, and other norms of social justice; as part of a complex web of interdependent rights, it transforms the struggle for ecological wellness in the common interest into a struggle for human dignity. As international environmental law scholars Birnie, Boyle, and Redgwell write, a rights-based approach recognises 'the vital character of the environment as a basic condition of life, indispensable to the promotion of human dignity and welfare, and to the fulfilment of other human rights'.[58] A rights-based approach to ecological governance thus raises the stakes against those who would damage our natural world.

Third, human rights serve as crucibles for human security and democracy. The security of the human person individually and collectively (e.g., 'freedom from fear', 'freedom from want') is more or less synonymous with the fulfilment of human rights. As US President John F Kennedy once famously asked, '[i]s not peace, in the last analysis, basically a

---

[56] R Dworkin, *Taking Rights Seriously* (Cambridge MA, Harvard University Press 1977) 91–3, 189–91, 269.

[57] MS McDougal, 'Perspectives for an International Law of Human Dignity', in MS McDougal et al., *Studies in World Public Order* (New Haven CT, Yale University Press 1987; Dordrecht, Martinus Nijhoff 1987) 987, 987–1019.

[58] PW Birnie, AE Boyle and C Redgwell, *International Law and the Environment* (3rd edn, Oxford and New York, Oxford University Press 2009) 278–79.

matter of human rights, the right to live out our lives without fear of devastation – the right to breathe air as nature provided it – the right of future generations to a healthy existence?'[59] But a reverse logic is likewise true: to fulfil human rights is to promote peace. Respect for others – mutual respect – is the foundation of peace and human security; it also is at the innermost core of all human rights. In virtually any societal context, taking human rights seriously, committing to mutual tolerance and reciprocal forbearance, even making only a good faith effort to do this, is more likely than other stratagems to bring peace and tranquillity to social order.

This is particularly true when contemplating the life-enhancing and life-sustaining qualities of Mother Nature. Citing the linkages between basic needs and environmental well-being, James Quilligan observes: 'The basic reason for alleviating material insecurity – through food, clean water, housing, health care, education, jobs and self-sustaining livelihoods – is to ensure people's personal safety and survival in conditions of peace and dignity'.[60]

Fourth, as markers of societal values, agents of human dignity, and bases of human security, human rights challenge State sovereignty and power and the particularist agendas of private elites. On countless occasions, states, intergovernmental institutions (IGOs), nongovernmental organisations (NGOs), professional associations, business enterprises, trade unions, faith-based groups, and others have relied on this 'corpus juris of social justice'[61] to measure and curb State misbehaviour. Today, human rights law and policy are increasingly invoked to ensure that private corporate behaviour – which tends to be more powerful than the political jurisdictions within which corporations operate – is made accountable to the common good. Human rights enshrine an entitlement to equal, non-discriminatory treatment that belongs to all human beings everywhere, transnational settings included. As such, they place significant moral, legal, and political burdens on those who treat other human beings and their basic needs – environmental and otherwise – in disrespectful, prejudicial ways. The potential of human rights to dislodge

---

[59] JF Kennedy, *A Strategy of Peace* (Commencement Address at American University, 10 June 1963), http://www.jfklibrary.org/Asset-Viewer/BWC7I4C9QUmLG9J6I8oy8w.aspx (accessed 1 October 2014).

[60] JB Quilligan, 'Commons for Peace' (Fall/Winter 2011) *Kosmos* 15, 15–16.

[61] C Van Boven, 'Survey of the Positive Law of Human Rights', in K Vasak (ed.) *The International Dimensions of Human Rights, Vol. 1* (Westport CT, Greenwood Press and UNESCO, Paris; revised and edited for the English transl. by Philip Alston 1982) 85–110, 88.

or seriously burden State and private interests that commit and perpetuate environmental abuse, a growing trend, is compelling.

Finally, by virtue of the above four attributes, human rights denote legal and political empowerment. They confer a sense of entitlement on the part of the rights-holder 'to know and act upon his [sic] rights'.[62] They impose a corollary duty on the part of the State (and others in positions of authority and control) to provide satisfaction or redress to the extent warranted by the facts involved. The essence of human rights law and policy is that, in Michael Freeman's pointed alert, 'if you have a right to x, and you do not get x, this is not only a wrong, but it is a wrong against *you*'[63] – a truth that extends inexorably to environmental rights-holders (living and unborn, principal or surrogate) and validates Cicero's great insight that 'freedom is participation in power'.[64] Because human rights entail fundamental values of superior order, their violation correspondingly entails greater moral and legal condemnation than do other wrongs and therefore greater levels of accountability, in some cases even criminal liability. Because they help shift moral and legal burdens, they redistribute power and thereby facilitate new grounds for expression through which '[t]he more fortunate are called upon to assist the less fortunate as an internationally recognized responsibility'.[65] And because they afford a growing, more inclusively responsive jurisprudence upon which aggrieved claimants may stand, so also do they provide increasing access to institutions (national and international; formal and informal)

---

62 Quoting Paragraph 7 of the 1975 Helsinki Accords, officially known as the Final Act of the Conference on Security and Co-operation in Europe: Declaration on Principles Guiding Relations Between Participating States, Respect for Human Rights and Fundamental Freedoms, Including Freedom of Thought, Conscience, and Religion or Belief (1 Aug 1975), reprinted in 14 *Int'l Leg. Mats* 1292 (1975) and BH Weston and JC Carlson (eds) I *Int'l L. & World Order: Basic Documents* I.D.10 (1994–), http://nijhoffonline.nl/subject?id=ILWO (accessed 1 October 2014).

63 MDA Freeman, *Human Rights: An Interdisciplinary Approach* (Cambridge, Polity Press and Malden MA, Blackwell 2002) 61 (emphasis added).

64 As quoted in *Marcus Tullius Cicero Quotes, Goodreads*, http://www.goodreads.com/quotes/show/106593 (accessed 1 October 2014).

65 M Robinson, 'Foreword', in M Santos Pais, *A Human Rights Conceptual Framework for UNICEF* (Innocenti Essay No. 9, UNICEF International Child Development Centre, Florence 1999), http://www.google.com/url?sa=t&rct=j&q=&esrc=s&source=web&cd=1&ved=0CB4QFjAA&url=http%3A%2F%2Fdireitoshumanos.gddc.pt%2Fpdf%2Fessay-9.pdf&ei=EOD8U4r5Js2qyASdrIGICA&usg=AFQjCNHlBPBPHTpQLQEsYwoOUm_NnwBoWQ&sig2=as-aPJJETp9xJO79zEl_Tw&bvm=bv.73612305,d.aWw (accessed 1 October 2014).

through which their human rights claims may be vindicated in whole or in part. At a time when conventional representative democracy, not just autocracy, is showing itself to be appallingly dysfunctional, human rights offer a way to regenerate and reignite democratic process in legal and political decision-making, critical to addressing our portentous environmental future.

In sum, human rights law and policy provide a way of life – a latter day Golden Rule – with which no other known ethical system can compare. It is, thus, for abundant good reason that the UN Human Rights Council's first Independent Expert on the Issue of Human Rights Obligations Relating to the Enjoyment of a Safe, Clean, Healthy and Sustainable Environment recently stated that 'human rights have become the *lingua franca* of ethical discourse'.[66]

It makes sense, too, that human rights should be the *lingua franca* of commons-based ecological governance. Commons are, after all, human creations, and, like all other human systems (the State and Market included), they are fallible, vulnerable to abuse (e.g., black markets, inner-city gang operations, Internet pirates, and other criminal arrangements). There is, thus, every reason that the values of liberty, equality, and community should discipline thc internal governance of commons as much as any other polity. Indeed, if commons-based ecological governance is to cultivate trust and therefore its chances for stable management, its constitutional and operational rules – its Vernacular Law – must be driven by an ethic of respect for nature, sufficiency, interdependence, shared responsibility, and fairness, and by a logic of integrated global and local citizenship that insists upon transparency and accountability in all activities that affect the integrity of the environment. These values are deeply embedded in human rights law and policy, especially since the advent of 'nature's rights' in recent years and the essence of which consists of 'rescuing "*the right to existence*" of human beings themselves [who] cannot live apart from Nature', in the words of Alberto Acosta, former President of Ecuador's Constituent Assembly.[67]

---

[66] J Knox, 'Climate Ethics and Human Rights' (2014) 5 *J. Hum. Rts. & the Env't* 22–34, 23. Accord, M Ignatieff, *Human Rights as Politics and Idolatry* (Princeton NJ, Princeton University Press 2001) 53 ('Human rights has [sic] become… the *lingua franca* of global moral thought').

[67] A Acosta, *Toward a Universal Declaration of Rights of Nature* (14 Aug 2010) (article for the AFESE Journal), in unpaginated manuscript form at https://www.google.com/?gws_rd=ssl#q=Alberto+Acosta%2C+Toward+a+Universal+Declaration+of+Rights+of+Nature+ (accessed 1 October 2014).

Hence our advocacy not just of commons-based ecological governance, but of commons- *and right-based* ecological governance (i.e., 'green governance') – and, further still, of a human right *to* green governance.

The importance of a human right to green governance does not end here. Besides helping to ensure that human rights inform the internal administration of commons governance, it is a powerful means for enabling green governance itself. This, too, we advocate for good reason. The human right to a clean and healthy environment[68] as currently conceived and practised is fractured between three different conceptions: as a *substantive entitlement* autonomous unto itself; as an *entitlement derived* from other substantive rights, and as a cluster of *procedural entitlements*. Each conception relies upon different sources of legal authority, many of which are 'soft law' expressions of environmental protection. Each on its own terms, however robust in their immediate dominions, is limited in its official recognition and jurisdictional reach. And together relative to the popular, overarching right to environment concept, they suggest a failure of definitional consensus, a condition to which jurists ascribe, often appropriately, the notion of 'indeterminacy', a quality of vagueness that can serve as grounds for 'nonjusticiability' (i.e., claim or case dismissal).

But like many evasive judicial manoeuvres, the argument that the formulation 'human right to a clean and healthy environment' is too indeterminate to be applied by a court of law is open to challenge. Common law courts in India, Nepal, Pakistan, and elsewhere in South Asia, for example, have had little or no difficulty in recognising and applying it. The more likely explanation, we therefore submit, lies elsewhere: in the legal and political resistance of the world's dominant economies where repeatedly the utility and carrying capacity of natural systems have been outweighed by 'free market' values.[69] As long as contemporary ecological governance remains in the grip of essentially unregulated or neoliberal capitalism (what social critic Naomi Klein calls

[68] More precisely, a 'human right to a clean, healthy, biodiverse, and sustainable environment'; but hereinafter, for convenience, 'a human right to environment' or 'right to environment'.

[69] It must be added that neither Soviet-based communism nor Chinese-style State capitalism has shown an ability to transcend this way of thinking and governing any more than the capitalist West.

'disaster capitalism'),[70] there never will be a human right to environment widely recognised and honoured across the globe in any formal/official sense.

These are problems that, for the most part, do not beset a human right to green governance. First, green governance shares with the standard right to environment a worldview and sensibility that is ecumenical in spirit and analysis. But most notably, green governance offers five compelling comparative advantages (some of them already familiar). It is anchored in a venerable legal history that stretches back to the Roman Empire and the Magna Carta and its companion Charter of the Forest, thus providing impressive sources of instruction, credibility, and guidance for political and legal innovation. It grows out of a serious intellectual framework and discourse that leads to quite specific principles of internal commons governance. It provides a penetrating critique of Market culture, validates human cooperation and community, and otherwise emancipates us from *homo economicus* models of human existence, thus avoiding the epistemological flaws and systemic limitations of neoliberal ideology. It offers access to a rich array of precedents in the form of successful working models for provisioning and empowerment, which in many instances out-perform the State/Market, thus providing positive, constructive alternatives that help decision-makers reach common-sense, eco-sensitive results. Finally, it honours the best from all political ideologies, embracing the importance of community and fiscal responsibility (conservatism), individual initiative (libertarianism), equality and basic needs (liberalism), and constraints on markets (leftism). In this way, green governance provides a rudimentary basis for new forms of trans-ideological cooperation and consensus.

In sum, taken seriously, the human right to green governance is not only relatively free of indeterminacy but gives us many reasons for believing that widespread commons- and rights-based ecological governance can be a future reality. It affirms what Sam Adelman would likely call a 'meta right' – 'a foundational right that would, where necessary, take precedence over other rights'; or alternatively, given the problematic of 'a hierarchy at odds with the assertion that all rights are equal and indivisible', a 'species right' ('a new category of … right which transcends traditional categorizations, highlights the truly universal nature of the [environmental] threat, and which we hold not simply as individual human beings but rather by virtue of our membership [in] the

---

[70] N Klein, *The Shock Doctrine: The Rise of Disaster Capitalism* (New York, Metropolitan Books/Henry Holt 2007).

species *homo sapiens* …').[71] As such, it holds out the very real prospect of being able to mobilise and energise imagination, responsibility, and powerful synergies in support of green governance worldwide.

Of course, merely touting the virtues of a human right to commons- and rights-based ecological governance as a means for achieving it is not enough. For reasons stated at the outset of this chapter, it would be naïve to expect that the State and Market would easily accept a supportive role in its recognition and enforcement. But already we are seeing, as outlined above, a growing international movement of self-identified commoners and their convergence with allied movements such as the Social and Solidarity Economy, Degrowth, Transition Towns, co-operatives, peer production, and the 'sharing economy'.[72] As it becomes increasingly clear that the neoliberal project of ever-expanding consumption on a global scale is a utopian, totalistic dream, we believe the appeal of a human right to green governance – and its feasibility as a new legal stratagem – will become very real.

But, finally, a caution. There is today a dark side to human rights that poses a threat to all human rights claims or demands that fall outside the boundaries of civil and political rights, the human right to green governance included. With the end of the Cold War and the subsequent massive globalisation of capital according to neoliberal rules of production and consumption, all other rights (economic, social, cultural, and communitarian or group rights) have been increasingly marginalised as 'aspirations' rather than being given full recognition as 'rights'.[73] As Tony Evans observes, 'the neoliberal consensus upon which the practices of globalization are built has succeeded in establishing the language of civil and political rights as the acceptable voice – indeed the only legitimate voice – of human rights talk'.[74] In fact, with help from the

---

[71] S Adelman, 'Rethinking Human Rights: The Impact of Climate Change on the Dominant Discourse', in S Humphreys (ed.), *Human Rights and Climate Change* (Cambridge UK, Cambridge University Press 2010) 159–80, 173. Out of respect for all life on Earth, however, it would be helpful to view the species *homo sapiens* living among all other species.

[72] The Degrowth Conference in Leipzig, Germany, in September 2014, is a good example of this convergence. Some 2,700 people from all of these movements convened in more than 200 panels to explore new modes of collaboration on commons-based strategies that could address environmental harms, social injustices, the provisioning of basic needs, and relentless economic growth. See http://leipzig.degrowth.org/en (accessed 1 October 2014).

[73] T Evans, 'Citizenship and Human Rights in the Age of Globalization' (2000) 25 *Alternatives: Global, Local, Political* 415–38, 416–22.

[74] Ibid at 416.

courts, civil and political rights – which presumably are intended for natural persons – are today being engineered to benefit corporations,[75] to help them claim the protection of their favoured human rights without having to accept responsibility for their violations of economic, social, and cultural rights, and communitarian or group rights which generally they disfavour.[76]

The human right to green governance, no less than the standard human right to environment, is at risk in this respect. The collective nature of commons, which runs against the grain of the narrow neoliberal conception of human rights, makes this all but inevitable. Indeed, some fear the survival of even the ecumenical Universal Declaration of Human Rights.[77] But the human right to green governance, with its many comparative advantages, seems clearly to be the most capable of withstanding this challenge. For this, however, widespread support is mandatory. So to this end we offer a Universal Covenant Affirming a Human Right to Commons- and Rights-based Governance of Earth's Natural Wealth and Resources, an assertive statement that we hope will catalyse positive energies to explore and promote the merits of green governance at all levels of social organisation.[78]

## 7. THE CHALLENGE OF MULTILEVEL, COMPLEXITY-BASED GOVERNANCE

Arguably the greatest challenge in making green governance a reality is imagining how to induce the dominant State/Market order, nationally and internationally, to recognise and foster commons- and rights-based ecological governance. To this end, all the skills and tools known to effective

---

[75] See e.g., *Citizens United v Federal Election Commission*, 558 U.S. 310 (2010). See also, M Emberland, *The Human Rights of Companies: Exploring the Structure of ECHR Protection* (Oxford, Oxford University Press 2006), exploring the jurisprudence of the European Court of Human Rights.

[76] For insightful detail and analysis, see A Grear, 'Corporations, Human Rights, and the Age of Globalization: Another Look at the "Dark Side" in the Twenty-First Century', in BH Weston and A Grear (eds), *Human Rights in the World Community: Issues and Action* (Philadelphia, University of Pennsylvania Press, forthcoming in 2015).

[77] Ibid.

[78] Available at (Sept 2013) 4/2 *J. Hum. Rts. & Env't* 215–25; also at Commons Law Project, http://commonslawproject.org (accessed 1 October 2014).

persuasion must be brought to bear, and at all levels of social organisation – while also maintaining a healthy realism about entrenched political systems. No amount of persuasion is likely to yield progress if it is focused exclusively or even primarily on the formal legal order and its array of sanctioned political action. The 'top down' strategies favoured by guardians of the current order will tend to reflect the interests of the 'haves' over the 'have nots' and a fierce commitment to preserving the existing order. Or it will seek to co-opt the quest for fundamental change.

Without dismissing appropriate 'top-down' initiatives altogether, we believe the greatest promise lies in 'bottom-up' or grassroots-driven approaches, especially those that are inclusive and cross-sectoral. This is not just a political opinion: profound discoveries in the evolutionary sciences and the rise of complexity science over the past generation validate the power of bottom-up forms of social organisation and governance. Extensive empirical research shows that some of the most robust, stable forms of governance are distributed, self-organised, and collaborative.[79] It is thus important to survey this field, even if briefly, because these scientific fields point to a different framework for understanding human agency, the evolution of cooperation, and the dynamics of governance in a networked environment.

The worldview that has prevailed for the past several centuries is familiar but archaic. As we have previously noted, it sees humanity as separate from Nature, and posits a fairly static, mechanical worldview in which knowable causes produce measurable effects in linear patterns. In this mindset, the point is to improve the rigour of our instruments and empirical analysis so that we might identify cause-and-effect more clearly, and then regulate and control isolated elements of nature or human society. Within this mindset, governance is focused on amassing the most extensive fact base and objective expert knowledge so that political and scientific elites can devise more reliable (usually bureaucratic) systems for achieving desired results.

---

[79] See n 17 and accompanying text on various commentators on self-organised, bottom-up forms of cooperative governance. It is also worth consulting ED Beinhocker, *The Origin of Wealth: Evolution, Complexity and the Radical Remaking of Economics* (Boston, Harvard Business School Press 2006) for his insights into complexity theory and its logic. The sub-discipline of holonics is also germane here. See M Ulieru, 'Organic Governance Through the Logic of Holonic Systems', in *From Bitcoin to Burning Man and Beyond: The Quest for Identity and Autonomy in a Digital Society* (Boston MA, ID3 2014), https://idcubed.org/chapter-11-organic-governance-logic-holonic-systems (accessed 1 October 2014).

Complexity science has opened the door to some quite different frameworks for understanding human and ecological phenomena. The field draws on the lessons of evolution, chemistry, and biology to identify fundamental principles governing what it calls 'complex adaptive systems'. Frequently cited complex adaptive systems include the self-organising dynamics and behaviour of the brain, cells, ant colonies, the biosphere, the stock market, and Internet communities. Remarkable parallels can be traced between the behaviours of natural, physical systems and social and economic systems.

Complexity science proposes a conceptual framing of the world based on the observable behaviour of complex natural systems. Stable, successful systems are not constructed from advance blueprints devised by brilliant minds (e.g., God as the absent watchmaker); they can be shown to have self-organised from the free interplay of adaptive agents following simple principles operating at the local level, with no big-picture knowledge or teleological goals at the outset. Instead of presuming that an *a priori*, comprehensive design system is possible and can yield the best outcomes, complexity theory takes its cues from bio-physical evolution. It asserts that the best results will arise if intelligent, living agents are allowed to evolve over time towards optimum outcomes in suitable environments. The schemas or agents that survive and thrive will be the ones capable of prevailing against competitors and reproducing; less capable agents will be shunted to niches or die, according to principles of natural selection.

Microbes, ants, humans, and diverse other organisms exhibit characteristics of complex adaptive systems. Each is nested within larger complex systems that are dynamic and constantly shifting; and yet each flourishes by embodying some highly predictive theories, as distilled in schema that are useful in exploring resources and regularities in a particular environment (the 'fitness landscape'). The species with the most adaptive schema (e.g., DNA, organism functions, culture) and the most refined feedback loops will be better equipped to learn from its environment and thus adapt, evolve, and grow. Evolutionary scientists increasingly believe that natural selection manifests itself more at the group level than through individual organisms.

What do these bodies of thought have to do with our search for green governance? They suggest that human communities can evolve into higher, more complex forms of organisation without the directive control of a central sovereign or bureaucracy. Given a sufficiently defined and hospitable fitness landscape, self-organisation based on local circumstances can occur. Just as biological and chemical systems exhibit autocatalytic features that generate 'order for free', so human communities have inborn capacities to create stable order. Indeed, this is one of the key insights of Nobel laureate Elinor Ostrom's empirical research of natural resource commons, and the countless self-organised communities on the Internet constitute a kind of existence proof. The capacity for self-organisation is often overlooked, especially by mainstream economics and its model of *homo economicus*.[80]

Yet commons are fully capable of yielding robust, flexible, and durable forms of managing ecological resources. These advantages stem in part from the fact that commons governance systems arise organically from the governed themselves in ways that are mindful of the resources, local conditions, and cultural norms; externally imposed or arbitrary systems may or may not be as adaptive.

The point is that effective governance need not be imposed through a comprehensive grid of uniform general rules embodied in formal State Law and administered by centralised legislatures, regulators, and courts. Complexity science demonstrates that governance can be a highly distributed, evolving form of social practice and tradition; and it can arise from Vernacular Law that is rooted in communities of decentralised agents responding to particular local circumstances. The twentieth-century mind may have been convinced that governance and organisation must be based on uniform, top-down expertise and command, but the lessons of evolutionary sciences and complexity science suggest that new modes of diversified, locally appropriate governance are entirely feasible, based on what we know about bio-physical and social systems.

The lessons of complexity theory hold clues to how we might evolve new forms of multilateral governance that transcend the entrenched governance problems of the State as allied with neoliberal Market capitalism. The implications for commons- and rights-based governance are enormous. A central problem with existing international environmental and human rights law – the human right to environment included – is its dependence on a territorially based, consensual system of global governance in which rigid State sovereignties are empowered, essentially alone, to make the legal and political decisions about problems that

---

[80] An excellent overview of economics as a complex adaptive system can be found in ED Beinhocker, *The Origin of Wealth: Evolution, Complexity and the Radical Remaking of Economics* (Boston MA, Harvard Business School Press 2006). Two useful popular accounts include M Gell-Mann, *The Quark and the Jaguar* (New York, W.H. Freeman & Co. 1994) and M Mitchell Waldrop, *Complexity: The Emerging Science at the Edge of Order and Chaos* (New York, Simon & Schuster 1992).

international environmental and human rights law are supposed to solve. Another problem is that, however high-minded their rhetoric, sovereign States typically act tenaciously in their own self-interest, generally perceived in neoliberal economic terms with little to no regard for the ecological and social needs of the wider community of which they are a part.

If, however, our planet is to survive in a manner truly hospitable to life on it, international environmental and human rights law – indeed, national and international law in general – must change. Law as a creature of sovereign hierarchies must adapt to a world of interdependent, interpenetrating networks, both ecological and social. Over the past twenty years, the Internet has significantly dissolved institutional and geographic boundaries, making them far more porous, if not indefensible. This poses serious challenges to conventional forms of law, not just in terms of geographic control, but equally in terms of maintaining legitimacy and efficacy. The transnational character of the Internet and the social cooperation that it is fostering, is opening up new opportunities to reimagine standard forms of law, which are often too remote and detached from on-the-ground moral, socioeconomic, political, and cultural realities. We have already seen how many nimble Internet communities have outflanked – at least provisionally – stodgy government bureaucracies and politicians (e.g., the Arab Spring; citizen protests in Madrid, Istanbul, London and Athens; the WikiLeaks and Edward Snowden disclosures of US government documents).

When law is controlled mostly by elite policymakers and administered through arcane and often corrupt legal systems, even as open networks enable a divergent public narrative to emerge, citizens understandably become cynical about formal law. They are not truly agents of its creation and interpretation, even in nominal democracies. And when they do become engaged, perhaps as a result of crisis or demagoguery, there may not be vehicles for implementing governmental laws and policies. Elinor Ostrom put it this way on winning the Nobel Prize:

> I'm not denigrating that officials can do something very positive, but what we have ignored is what citizens can do, and the importance of real involvement by the people involved, as opposed to just having somebody in Washington or at a far, far distance, make a rule. How does that get all the way down to management of forests, fisheries, irrigation systems, etc.? So we have to look from the ground up.[81]

---

[81] Professor Elinor Ostrom, Press conference at Indiana University, following the announcement of her shared award of the 2009 Nobel Prize in Economic

A new system of multilateral ecological governance, therefore, must reimagine the role of the State and multilateral institutions and their policy priorities. This requires the development of a new vision of societal development beyond maximum capital accumulation and economic growth. A vision of the 'good life' beyond material acquisition and an economics of sufficiency must be developed. This will of course entail both profound shifts in all aspects of life over many years and realignments of international political relationships. The State must move from serving as the sovereign master of a closed, hierarchical system, to the light-touch host of an open, diverse network.

In other words, there must be means for the agency of commoners at lower levels of governance to be expressed rapidly, dynamically, and interactively at the higher macro-levels levels of law. This capacity is important for the quality of information, the flexibility and speed of response, and the overall legitimacy of international governance, ecological and otherwise. A crude model for such transnational, networked collaboration is the rapid response of health researchers and public health agencies to the outbreak of the severe acute respiratory system (SARS) virus in 2002.[82] Networked collaboration was widely seen as essential to the quick containment of an infectious disease that otherwise had catastrophic potential; one can only wish that such fierce cooperation could animate efforts to reduce carbon emissions into the atmosphere.

If online networks could knit together diverse commoners and make them a visible and coordinated political force in international legal fora, it would become possible to make law conform more closely to vernacular morality and practice and to change the political climate for new initiatives. In ecological design, there is a term to describe how natural phenomena at different spatial scales interconnect. 'Nature's processes

---

Sciences (carried live via satellite and streamed live on the Internet at http://www.nobelprize.org/mediaplayer/index.php?id=1223, 12 Oct 2009).

[82] During the SARS outbreak, the World Health Organization regularly convened teleconference meetings that allowed more than a dozen national public health agencies and leading medical laboratories to share information with each other rapidly. This collaborative approach enabled medical authorities to identify the virus and develop diagnostic tests and treatment regimes in a matter of weeks, not months, as would have been required had national health systems worked independently of each other. There are obvious analogous lessons for environmental initiatives.

are inherently *scale linking*, for they intimately depend on the flow of energy and materials across scales', write Sim van der Ryn and Stuart Cowan.[83]

> The waste oxygen from blue-green algae is absorbed by a blue whale, whose own waste carbon dioxide feeds an oak tree. Global cycles link organisms together in a highly effective recycling system crossing about seventeen tenfold jumps in scale, from a ten-billionth of a meter (the scale of photosynthesis) to ten thousand kilometers (the scale of the Earth itself ).[84]

Taken as a metaphor, if not a functional template, scale linking must find expression in national and international law to overcome the structural frictions and missing feedback loops that make existing institutions so ineffective. The Vernacular Law of commons at all levels needs to be integrated into the actual formulations of national and international law. The obvious vehicle for reimagining new types of scale-linked multilateral institutions is the Internet. Software platforms are just one channel among many in political relationships, and they are constitutive in the ways that could help reimagine more participatory, constructive modes of multilateral governance. By allowing information and participation to flow from the bottom up, from diverse levels and locations, the overall system of governance would be more capable of addressing the myriad, distributed complexities of ecosystem problems. The principles of polycentrism and subsidiarity could be designed into such a system.

Admittedly, what we propose is a general concept, not an implementation. But the virtue of such a system of networked multilateral governance is that it could help us get beyond the dysfunctional premises of the current system. It could provide new platforms for commoners to enact their ecological and human rights interests, and thereby help align governance more closely with ecological (and social) needs.

## 8. CONCLUSION

It is clear that any successful effort to deal with our multiple, cascading array of ecological harms must begin to address the deep systemic roots of the global environmental *problematique.* It is no longer credible to propose piecemeal or policy-specific 'solutions' for problems that are insinuated into the deep structures of global commerce and cultural

---

83 S van der Ryn and S Cowan, *Ecological Design* (Washington DC, Island Press 2007) 36.

84 Ibid.

habits of everyday life. What is most needed are new approaches that propose a new 'constitutional order' for commerce, social organization, and culture at all sorts of 'micro' and 'intermediate' levels, culminating in the future in new sorts of 'macro-scale', transnational governance systems.

Green governance provides us with a hardy imaginary for just such a reconstruction. It is not a focused 'solution' that will have cause-and-effect results, but rather a reimagining of the socioeconomic, political, and legal landscape in ways that, consistent with an inclusive understanding of human rights, invites broad empowerment, cooperation, transparency, innovation, and fairness. This is arguably the only way that societies will be able to marshal the resources needed to confront our many ecological crises. By understanding, too, that this is a profound human rights struggle – a struggle to build new and humane governance frameworks that will be more effective than the existing order allows – we can begin to build a stronger foundation for enlarging human rights, like the Commons, as a lived reality.

# 10. Towards new legal futures? In search of renewing foundations

**Anna Grear**

## 1. INTRODUCTION

This chapter is a development of an article originally published as part of an edition in the *Oñati Socio-Legal Series*.[1] The structural organisation of the original article traced the structure of an Oñati workshop held in 2012, offering a contingent weaving[2] of selected themes suggested by the ebb and flow of notes taken during discussions.[3]

[1] A Grear, 'Towards a New Horizon: In Search of a New Social and Juridical Imaginary' in A Grear (ed.), *Human Rights and the Environment: In Search of a New Relationship* (2013) 3/5 *Oñati Socio-Legal Series* 966–90. I would like to thank the anonymous referees and the editorial team of the *Oñati Socio-Legal Series* for their invaluable help and critique, as well as to thank Lorraine Code and Andreas Philippopolos-Mihalopoulos for their generous readings of that earlier work. I am also grateful to the Oñati International Institute for the Sociology of Law for awarding the original workshop, 'Human Rights and the Environment: In Search of a New Relationship', held there in July 2012.

[2] This 'provisional weaving' is fully aware of the incompleteness, indebtedness and limitations of any academic contribution, and offered in the spirit of Deleuze and Guattari's reflection on the nature of any book: 'A book has neither object nor subject; it is made of variously formed matters, and very different dates and speeds. To attribute the book to a subject is to overlook this working of matters, and the exteriority of their relations' G Deleuze and F Guatarri, *A Thousand Plateaus* (London and New York, Continuum 2004) at 4.

[3] I am especially grateful to Louis Kotze, Karen Morrow and Evadne Grant for a set of rough notes taken at the conference from which the present chapter is formed as a kind of reflection upon their contents in 'conversation' with my own readings in a wider range of scholarship. Now, see E Grant, LJ Kotze and KL Morrow, 'Human Rights and the Environment – In Search of a New Relationship: Synergies and Common Themes' (2013) 3/5 *Oñati Socio-Legal Series* 953–65.

This chapter follows the same broad strategy. Its animating purpose is to reflect upon insights yielded by a lively confluence between critical environmental law scholarship and new materialist accounts (from which critical environmental law itself draws), placing these in direct engagement with a search for a renewing socio-juridical imaginary[4] in which the relationship between human rights and the environment might be re-imagined. The chapter begins by contextualising its concerns in the semiotic and material dilemmas presented by the emergence of Anthropocene discourse, especially concerning the construction of a new universal humanity, before examining the shortfall of law's response to the climate crisis – and exploring some broad directions for a new juridical imaginary holding out the promise of new human and environmental foundations for law and legal processes.

## 1.1 Framing Our Reflection: Of Anthropocene Subjectivity

It would be difficult to overstate either the urgency or the complexity of the current epochal juncture in which countless human beings now struggle to avert the 'future histories' threatened by neoliberal globalised capitalism and climate crisis in the Anthropocene age. The planet itself, it is said, has entered an epoch in which 'humanity' itself has emerged as a geological force (*anthropos* ('human being/man') and *kainos* ('new'/ current')),[5] as a result of the 'large-scale human modification of the Earth System, primarily in the form of climate change, the most salient and perilous transgression of Holocene parameters'.[6] The human species, it is argued, has propelled itself to full spectrum dominance.[7] And it has done so with such deadly planetary implications that 'humanity' itself (despite histories of inequality and the patterned distributions of life, death, vulnerability and resilience marking the 'human' journey) now stands

---

4 The use of the term 'imaginary' is here intended to evoke the ways in which liberal law functions as a materio-semiotic and concrete manifestation of a dominant worldview, now operating globally as a frame for sense-making and action. The term invokes a socio-juridical *habitus* in and through which, arguably, juridical formations do some of their most powerful work as constructs for the 'normalisation' of a particular, dominant construction of 'reality'.

5 PJ Crutzen, 'Geology of Mankind' (2002) 23 *Nature* 415.

6 A Malm and A Horborg, 'The Geology of Mankind? A Critique of the Anthropocene Narrative' (2014) 1/1 *The Anthropocene Review* 62–69 at 63.

7 Crutzen (n 5).

united – or so Chakrabarty claims – in a new negative human universality predicated upon common vulnerability in the face of climate crisis.[8]

Chakrabarty asks the fundamentally important question of whether it is fair to include the 'poor of the world', whose carbon footprint is 'small', in responsibility for the genesis of the Anthropocene,[9] but concludes that the climate crisis, unlike the crisis of capitalism, unites humanity in a new negative universal: 'Unlike in the crises of capitalism', Chakrabarty explains, 'there are *no lifeboats here for the rich and the privileged* (witness the drought in Australia or recent fires in the wealthy neighborhoods of California)'.[10] Chakrabarty's Anthropocene subject – while *not* the undifferentiated *species-agent* responsible for the genesis of the Anthropocene that mainstream accounts of the Anthropocene assume – is nonetheless a new human universal – forged by radical climate precarity, in the face of which even the rich and privileged are brought into a universality with the poor: 'we appear to have become one at the level of the species'.[11]

This new universal is born of the 'emergent, new universal history of humans that flashes up in the moment of danger that is climate change'.[12] It is, Chakrabarty insists:

> a question of a human collectivity, an us, pointing to a figure of the universal that escapes our capacity to experience the world. It is more like a universal that arises from a shared sense of a catastrophe. It calls for a global approach to politics without the myth of a global identity, for, unlike a Hegelian universal, it cannot subsume particularities. We may provisionally call it a 'negative universal history'.[13]

However, this new universal cannot escape the need for caution. In the words of Braidotti: '[I]s it not risky to accept the construction of a negative formation of humanity as a category that stretches to all human beings, *all other differences notwithstanding*?'[14]

This is an immensely important question, for Chakrabarty's Anthropocene universalism runs the risk of subsuming important hierarchical

---

[8] D Chakrabarty, 'The Climate of History: Four Theses' (2009) 35 *Critical Inquiry* 35: 197–222.

[9] Ibid at 217.

[10] Ibid 221, emphasis added.

[11] Ibid.

[12] Ibid.

[13] Ibid 222.

[14] R Braidotti, *The Posthuman* (Cambridge, Polity 2013) at 88, original emphasis.

patterns *within* Anthropocene climate precarity. Such patterns, moreover, express the radical unevennesses characterising not just the contemporary neoliberal world order but a long prior history of well-rehearsed practices of socio-economic privilege and exclusion implicated in the genesis of the climate crisis itself.[15] Malm and Hornborg, for example, argue that Chakrabarty's claim that there are 'no lifeboats here for the rich and privileged':

> blatantly overlooks the realities of differentiated vulnerability on all scales of human society: witness Katrina in black and white neighborhoods of New Orleans, or Sandy in Haiti and Manhattan, or sea level rise in Bangladesh and the Netherlands, or practically any other impact, direct or indirect, of climate change. For the foreseeable future – indeed, as long as there are human societies on Earth – there *will* be lifeboats for the rich and privileged. If climate change represents a form of apocalypse, it is not universal, but uneven and combined: the species is as much an abstraction at the end of the line as at the source.[16]

Malm and Hornborg are correct: it is vital to retain a critical engagement with the notion of the universal at the 'end of the line', just as much as 'at the source'. Well-practised distributions of privilege and disadvantage hinge upon past histories, but also presage a predictable future history of the self-same human hierarchies of being in the midst of Anthropocene precarity: Anthropocene vulnerability and resilience are unevenly distributed. This uneven distribution, moreover, reflects patterns visible in the genesis of the Anthropocene itself. Thus, while it seems relatively well established that the actors at the heart of the standard account of Anthropocene's industrial origins (and of the history of the neo-colonial international legal order) appear to be an identifiable, highly selective, dominant group of humans, present trajectories point to the need for a sustained engagement with the question of Anthropocene subjectivity in a refusal to universalise the situatedness of humans (and their 'non-human'

---

15 A Grear, 'Towards "Climate Justice"? A Critical Reflection on Legal Subjectivity and Climate *In*justice: Warning Signals, Patterned Hierarchies, Directions for Future Law and Policy' (2014) 5 *Journal of Human Rights and the Environment* 103–33.

16 A Malm and A Hornborg, 'The Geology of Mankind? A Critique of the Anthropocene Narrative' (2014) 1/1 *The Anthropocene Review* 62–69 at 63. Here the authors draw attention to A Malm, 'Sea Wall Politics: Uneven and Combined Protection of the Nile Delta Coastline in the Face of Sea Level Rise' (2013) 39 *Critical Sociology* 803–32; A Malm and S Esmailian, 'Ways In and Out of Vulnerability to Climate Change: Abandoning the Mubarak Project in the Northern Nile Delta Egypt' (2012) 45 *Antipode* 474–92.

'others') 'all other differences notwithstanding'.[17] Indeed, it seems urgent to theorise subjectivity (and the juridical imaginary – including juridical subjectivity) in a way capable of doing fuller *justice* to *all* the living, breathing entanglements of life on earth faced by climate crisis.

There is no doubt that the climate crisis is a 'game changer'. For some, indeed, it is the urgency of climate change and the peril faced by humanity and other living beings and systems that contextualises and partially explains the 'currency of posthumanist discourse on the non-human – the animal, the environment or the object – [suggesting] rising concerns about humanity's place in the ecological order'.[18] For a legal theorist, these concerns circle around the pressing question of how law – which operatively constructs 'humanity', the 'non-human', the 'animal', 'the' 'environment' and 'the object' – can respond to the need for a radical new imaginary capable of responding to the threat that co-situates us – in hierarchically mediated patterns – with countless 'other species' and living systems in conditions of *common – and uneven –* precarity.[19]

### 1.2 Of Law's Response

Law (including human rights law and environmental regulation and governance regimes) has quite simply – to date – failed to deliver the necessary response to ecological crises.[20] The track record is lamentable. International legal regulation and governance regimes exhibit profound limitations impeding genuinely transformative possibilities.[21] There seems to be a deepening of neoliberal capitalist ideology's grip on mainstream international strategies for addressing the eco-crisis: international law (itself an extension of Global North commitments and a history of colonialism)[22] seems committed to maintaining 'business as usual', while strategies for change remain captive to a reductive set of priorities and ideological structures directly linked to global warming.[23]

---

17 Braidotti (n 14).

18 F Chiew, 'Posthuman Ethics with Cary Wolfe and Karen Barad: Animal Compassion as Trans-Species Entanglement' (2014) 31 *Theory, Culture and Society* 51–69 at 51 (Abstract).

19 UNEP, GEO 5 Report, 2012.

20 K Bosselmann, 'Losing the Forest for the Trees: Environmental Reductionism in the Law' (2010) 2 *Sustainability* 2424–48.

21 S Turner, *A Global Environmental Right* (Routledge, London 2013).

22 A Anghie, *Imperialism, Sovereignty and the Making of International Law* (Cambridge, CUP 2005).

23 P Newell and M Patterson, *Climate Capitalism: Global Warming and the Transformation of the Global Economy* (Cambridge, CUP 2010); M Koch,

Developments such as the corporation-friendly failure of Rio+20 to deliver anything beyond market mono-culturalism;[24] the installation of discursive and legal strategies such as 'eco-system services' at the heart of environmental regulatory responses[25] (and examples of the so-called 'green economy' could be multiplied) signal, if anything, the *resistive* entrenchment of the neoliberal capitalist commitments so widely linked to climate destruction.[26]

Such commitments, moreover, are still complexly dominated by fundamentally Cartesian philosophical foundations,[27] expressed in the persisting ontological, epistemic, ethical and juridical primacy of 'human' centrality.[28] 'Nature', meanwhile – as Cartesian 'object' (mere 'matter') – is constructed not just as exploitable, but also as both priced and fungible: 'nature' and its 'resources' are merely interchangeable global commodity-elements within a hegemonic market system in which the human body is also thoroughly commodified – again, reductively conceived by processes for which Cartesian and Kantian subject-object relations are fundamental.[29]

The Anthropocene context presents the need for fresh engagements with juridical foundations. If climate precarity – to borrow Chakrabarty's phrase – 'flashes up' the need to imagine new, responsibilised ways of being, then it will be absolutely essential to replace the subject-object relations lying at the foundation of climate crisis and unjust patterns of the neoliberal international legal order alike.

The road to climate crisis (which should be understood as *a crisis of human hierarchy* (and of *human hierarchies of being*))[30] lies in the deeply eco-destructive implications of rationalist Western ontology and

---

*Capitalism and Climate Change: Theoretical Discussion, Historical Development and Policy Responses* (London and New York, Palgrave Macmillan 2012).

24 S Adelman,'Rio+20: Sustainable Injustice in a Time of Crisis' (2013) 4/1 *Journal of Human Rights and the Environment* 6–31.

25 For an invigorating set of engagements on this theme see (2014) 5/2 *Journal of Human Rights and the Environment* 107–219 (an edition devoted to 'ecosystem services and capitalism').

26 Koch (n 23).

27 L Code, *Ecological Thinking: The Politics of Epistemic Location* (Oxford, OUP 2006); Bosselmann (n 20).

28 Code, ibid; P Curry, *Ecological Ethics* (Cambridge, Polity 2006).

29 P Halewood, 'Law's Bodies: Disembodiment and the Structure of Liberal Property Rights' (1996) 1 *Iowa Law Review* 1331–93.

30 A Grear, 'Deconstructing *Anthropos*: A Critical Legal Reflection on "Anthropocentric" Law and Anthropocene "Humanity"' (forthcoming) (copy on file with author).

epistemology.[31] It is hardly surprising, therefore, now that we have reached the Anthropocene juncture, that there is an urgent need for a foundational re-imagination of *who 'we' are* in the 'world'.[32]

There is an urgent need for a renewing imaginary and set of social practices capable of giving birth to a new socio-ecological situation. How, though, might 'we' inaugurate a renewing imaginary to displace the stubborn destructiveness of the existing one? How might the Cartesian/Kantian subject and the subject-object relations so fundamental to rationalism, to capitalist exploitation and to climate degradation be rejected and replaced? Such questions have no easy answer. There are, however, some broad directions that suggest themselves.

## 2. BROAD DIRECTIONS FOR A NEW JURIDICAL IMAGINARY

Given the complexity of the current planetary predicament, and given a long and distinctively patterned history of privilege and predation associated with rationalistic subject-object relations, it seems especially important to diversify the search for new ways of being by drawing upon a wide range of constituencies and suitably diverse ecologies of insight. There is a particular need, in this regard, for a deep sensitivity not only to the complexities of the contemporary situation (with all its commonalities and uneveness) – but also for a profound and consciously chosen epistemic humility.

The present predicament – as posthumanist/new materialist approaches grasp so profoundly – requires both philosophy and law to move into a radical encounter with the irreducible complexity, movement, shifts and energies of living materiality itself. As will become clearer below, there is a related need to make *immanent* to legal epistemology a recognition that all our human descriptions of the living world, including our

---

[31] C Merchant, *The Death of Nature: Women, Ecology and the Scientific Revolution* (New York, Harper Collins 1980); V Plumwood, *Feminism and the Mastery of Nature* (London, Routledge 1993).

[32] L Code, 'Ecological Responsibilities: Which Trees? Where? Why?' in A Grear, *Should Trees Have Standing? 40 Years On* (Cheltenham, Elgar 2012) 84–99 at 92–99. Code seeks to present the question 'just who do we think we are?' in a distinctively provocative mode specifically designed to expose the false sense of entitlement implicit in the construct of the modern subject. Her question presents a timely, urgent and original political-ontological challenge entirely apt to the task of displacing the instituted imaginary with a new, ecological imaginary.

juridical categories, descriptors and tropes, along with attempts to co-ordinate relations in a living world, function within a greater complexity *irreducible* to linguistic or conceptual categories. Legal processes will need accordingly to embrace a new sense of *productive contingency* and be open to the destabilisation of habitual assumptions and practices in order to open law to dynamic insights emerging from the living, breathing engagements of the beings ('human' and otherwise) now *differentially* yet *commonly* situated in front of the current crises.

We begin, appropriately enough, with theoretical inquiry. However, a brief caveat is in order. Theory is misunderstood when it is seen as being a purely cerebral activity. Philosophising/theorising,[33] like all thinking, is inescapably an act of the embodied personality – a product of relations in a lived world from which the body is never separated.[34] Theoretical enquiry, in that sense, is always socio-haptic:[35] the philosopher/theorist, far from being the lonely Cartesian cogito with all-seeing-eye-from-nowhere (so beloved of Western intellectual myth) is always and everywhere an embodied being, formed socio-bio-eco-relationally. Thus, to 'begin' with philosophy/theory is not a choice to prioritise the cognitive or to elevate disembodied abstractions. 'Beginning' in this way – in the context of the present reflection – is merely to acknowledge the profound importance of reaching beyond the tired closures of the foundational philosophical commitments that have brought the world to its present human and ecological crisis/crises.

---

[33] There are various ways of conceptualising the relationship between philosophy and theory. At one level, philosophy is simply a mode of theoretical enquiry. However, there is also a sense in which philosophy and theory have been understood to be at odds: theory being associated with critical and resistive energies suspicious of the 'truth claims' of philosophy. Theory – in this register – can be most closely associated, perhaps, with the work of Derrida, Deleuze and Foucault, but also with earlier critical theory emerging from the Frankfurt School. For a stimulating discussion of the relationship between philosophy and theory, see J McCumber, 'Philosophy v Theory: Reshaping the Debate' (2008–9) 31 *Shibboleths: A Journal of Comparative Theory* 48–64.

[34] G Lakoff and M Johnson, *Philosophy in the Flesh: The Embodied Mind and its Challenge to Western Thought* (New York, Basic Books 1999).

[35] This term is intended to capture the sense in which even our most abstract concepts rely upon the use (both literally and virtually – or vicariously) of combinations of tactile environmental information, kinaesthetic analogies ('more' is 'up' and so forth – as Lakoff and Johnson (ibid) argue) and of our ingrained sense of personal and social situatedness.

## 2.1 Philosophy/Theory

Philosophy has always been a pivotal site for the re-imagination of the world, and it is important to give theoretical reflection its place in the search for a renewing socio-juridical imaginary. Theory – at its best – as bell hooks has observed, is a 'location for healing'.[36] Theory is the practice of thinking differently – of asking transgressive questions, of moving beyond sedimented patterns and of challenging the dominant order of thought in the world to reach for possible worlds otherwise. Theory understood in this resistive sense is thus precisely that which, in Rhadakrishnan's terms, 'states and elaborates the conditions of its non-acceptance of the world. [It] cannot be an acquiescence in the status quo … [the sole condition of its legitimacy is that it possesses an] ineluctable and dissenting worldliness'.[37]

Importantly, philosophy, and other forms of theoretical engagement with the central puzzles of existence, have contributed – and contribute – not just to everyday consciousness, but to everyday *un*consciousness (including the taken-for-granted socio-situational 'normal' inhabited and lived out in day-to-day bodily movement in and across multiple spheres).[38] Gatens has argued that 'philosophy … has much to answer for in relation to the character of the images and representations which dominate the everyday consciousness … philosophy [contributes – perhaps inevitably] to the cache of everyday consciousness'.[39]

Philosophy is foundational to the practices and images constituting the dominant modern socio-juridical imaginary. The dominant rationalist European philosophical suppositions informing science, markets, politics and law (including current international legal approaches to human rights and 'the environment') should be understood as *decisively influential* foundations of the road to eco-crisis.[40] These suppositions also function in a mundane way: they form the milieu – the collective habitus implicated by the devaluation and devastation of living creatures and

---

[36] b hooks, *Teaching to Transgress* (New York, Routledge 1995) at 59.

[37] R Radhakrishnan, *Theory in an Uneven World* (London, Blackwell 2003) at vi.

[38] For a particularly relevant reflection on situated movement and the life-world, see D Seamon, 'Lived Bodies, Place, and Phenomenology: Implications for Human Rights and Environmental Justice' (2013) 4/2 *Journal of Human Rights and the Environment* 143–66.

[39] M Gatens, *Imaginary Bodies: Ethics, Power and Corporeality* (London, Routledge 1996) at xii.

[40] Code (n 27); Curry (n 28); Merchant (n 31); Plumwood (n 31).

systems and the wholesale and the related oppression of multitudes of non-dominant human beings.[41] Western law has been pivotal in such processes, playing a central role in the transmission and reinforcement/construction of fundamentally Cartesian and Kantian commitments expressed in the centrality of a particular conception of the human (juridical) subject and the subject-object relations constructed with/through it.[42] A compelling body of scholarship now links the dominant philosophical foundations of Western law with globally salient forms of interlinked oppression functionally uniting non-dominant humans, animals and ecosystems in entangled oppressions.[43] It is clear – in a wide range of scholarship – that the current environmental crisis is ultimately legible as a crisis of human hierarchy decisively mediated by subject-object relations privileged by Cartesian and Kantian accounts of the person (and its 'environment').[44]

### 2.1.1 Ontology

> I have convinced myself that there is absolutely nothing in the world, no sky, no earth, no minds, no bodies. Does it now follow that I too do not exist? No: if I convinced myself of something then I certainly existed. But there is a deceiver of supreme power and cunning who is deliberately and constantly deceiving me. In that case I too undoubtedly exist, if he is deceiving me; and let him deceive me as much as he can, he will never bring it about that I am nothing so long as I think that I am something. So after considering

---

41 D Nibert, *Animal Rights, Human Rights: Entanglements of Oppression and Liberation* (Oxford, Rowman and Littlefield 2002); DM Ibrahim, 'A Return to Descartes: Property, Profit and the Corporate Ownership of Animals' (2007) 70 *Law and Contemporary Problems* 89–115.

42 Halewood (n 29).

43 Nibert (n 41); Ibrahim (n 41).

44 See HY Jung, 'Enlightenment and the Question of the Other: A Postmodern Audition' (2002) 24 *Human Studies* 297–306. Indeed, as Philippopoulos-Mihalopoulos points out, the etymology of the term 'environment' is French – deriving from '*environs*' which in turn derives from *en* ('in') and *virer* ('to turn'): 'This implies an inside that stands erect and an outside that surrounds this and turns around it. Environment is the "thing" that surrounds us, the dervish-like outside that whirls like a frilly skirt around a stable pivot … not only stable, fixed and unyielding but significantly "central"': A Philippopoulos-Mihalopoulos, *Law and Ecology: New Environmental Foundations* (Abingdon, Routledge 2011) at 22. Code deploys the term 'ecological' in order to avoid the implication of the environment surrounding or centring 'us'.

> everything very thoroughly, I must finally conclude that this proposition, I am, I exist, is necessarily true whenever it is put forward by me or conceived in my mind.[45]

As is well rehearsed in critical literatures, a core factor in the destructiveness of the Cartesian and Kantian philosophical imaginary is the centrality of a form of disembodied rationality – the shearing of the mind from both embodiment and emotions[46] – and a related objectification of the living order.[47] Unsurprisingly, perhaps, it is increasingly clear that any adequate response to the material exigencies and injustices characterising the twenty-first century requires an alternative ontology unambiguously embracing the radical centrality and inescapability of embodiment and its implications.

This insistence on the importance of the body, however, should not be misunderstood. This is not an insistence on a form of bio-essentialism, but rather a call to understand the meaning-making of materiality itself as a living field of communicative and agential significance. This is an approach calling for the body to be understood as an inseparable intertwining of bio-materiality and social construction thoroughly expressed in and through corporeality,[48] and for understanding corporeal materiality itself as an ethically and juridically significant feature of existence in its own right.

The present reflection begins with the centrality of human embodiment. This is precisely because the Cartesian and Kantian ontology and epistemology foundational to the dominant juridical imaginary radically separates the body from rationality (the latter being *the* defining characteristic of legal humanity). Beginning with the philosophy of Merleau-Ponty, the reflection will then move on to embrace the insights of Haraway and Barad.

In Merleau-Ponty's ontology, humans emerge as thoroughly intercarnal beings and the inherent sociality of human bodies emerges from the inescapable centrality of their feeling or perceptive faculties: there is

---

[45] R Descartes, *The Second Meditation* at 7:25.

[46] Halewood (n 29); A Bottomley, 'The Many Appearances of the Body in Feminist Scholarship' in A Bainham, SD Sclater and M Richards, *Body Lore and Laws* (Oxford, Hart 2002) 127–48; Gatens (n 39).

[47] Jung (n 44) Merchant (n 31); Plumwood (n 31).

[48] SJ Williams and G Bendelow, *The Lived Body: Sociological Themes, Embodied* Issues (London, Routledge 1998).

a sense in which the human body always reaches out (even pre-cognitively) towards a world from which it is never separated.[49] For Merleau-Ponty, the body is a living inter-corporeal flow: inescapably, humans participate in a continuing, animate reciprocation with the 'world' that our bodies perceive.[50] This flow takes place ceaselessly, and mostly far beyond the level of ordinary consciousness. The being-ness of the body, moreover, is always thoroughly *practical*, *sensory* and *involved*. Adams argues, importantly, that the body emerging from Merleau-Ponty's account is naturally '"ethesiological" – it is sensing, feeling, libidinal, erotic, desiring, empathic, and thus, as an immanent feature of its corporeal structure, "[t]he body asks for something other than … its relations with itself"'.[51] Such creaturely corporeality implies the dynamic *porosity* of materiality – bodies are a living sensory circuit with the world.[52] Indeed, in Merleau-Ponty's ontology, the human body is part of 'a single *reversible* fabric or *flesh*',[53] and the fundamental structure of *existence itself* is thoroughly *intercorporeal*.

Such radically interwoven ways of being signal possibilities as yet unreflected by our legal regimes, built as they are around the assumption of atomistic, separate subjects and objects. Consider the way in which law – and legal rights discourse – presupposes a unitary, disembodied self as the ontological foundation of the human legal subject:[54] singular, self-authorising, separative and bounded.[55] This is a hypostatisation completely at odds with the structure of existence as a dynamic, iterative

---

49 'The perceiving subject is not this absolute thinker; rather it functions according to a natal pact between our body and the world, between ourselves and our body', M Merleau-Ponty, 'An Unpublished Text by Maurice Merleau-Ponty: A Prospectus of his Work' in M Merleau-Ponty (JM Edie, ed.), *The Primacy of Perception* (Evanston IL, Northwestern University Press 1964) at 6.

50 M Merleau-Ponty, *The Phenomenology of Perception* (C Smith, trans) (Routledge Classics, London 2002). (First published by Routledge and Kegan Paul, 1962.)

51 WW Adams, 'The Primacy of Interrelating: Practicing Ecological Psychology with Buber, Levinas and Merleau-Ponty' (2007) 38 *Journal of Phenomenological* Psychology 24–61 at 40.

52 'Our own body is in the world as the heart is in the organism: it keeps the visible spectacle constantly alive, it breathes life into it and sustains it inwardly, and with it forms a system', Merleau-Ponty (n 50) at 235.

53 Williams and Bendelow (n 48) at 53.

54 A Grear, *Redirecting Human Rights: Facing the Challenge of Corporate Legal Humanity* (Basingstoke, Palgrave Macmillan 2010).

55 J Nedelsky, 'Law, Boundaries and the Bounded Self' (1990) 30 *Representations* 162–89.

folding – or 'chiasm'[56] – which is *intrinsic* to the structure of subjectivity and that of lived reality.[57] One implication of Merleau-Ponty's ontology is that 'subject' and 'object' are simply, as he puts it, 'two abstract "moments" of a unique structure which is *presence*'.[58] Indeed, a key implication of Merleau-Ponty's philosophy is that the very distinction between subject and object is rendered *inapt* – or more radically, ontologically *impossible.*

This is an insight given fresh potency by new materialist accounts of reality.

New materialist approaches are explicitly monist. They overtly reject Cartesian and related dualisms – and do so precisely by focusing on matter itself to produce a philosophy of radical immanence, drawing on the earlier work of philosophers and theorists such as Spinoza, Foucault, Lyotard, Deleuze, Guattari and Latour. New materialists adopt a broadly transversal approach, pushing difference to its limits in order to '[rewrite] all *possible and impossible* forms of emancipation'.[59]

While several contemporary new materialists can be identified as representatives of this vibrant philosophical approach, for reasons of space the present reflection will draw upon insights drawn from Barad and Haraway and then – turning to a more specific engagement with environmental law – that of Philippopoulos-Mihalopoulos, whose own monist philosophy of immanence draws directly upon Deleuze and can be regarded as an application of a new materialist ontology founding what Philippopoulos-Mihalopoulos calls 'a critical environmental law'.[60]

Turning first to the human body as a continuity between the work of Merleau-Ponty and a new materialist position, Merleau-Ponty's chiasmic flow between body and world is extended, for example, by the rich account of interwoven being provided by Haraway. In her account, the body itself is rendered a dynamic multiplicity:

> human genomes can be found in only about 10 percent of all the cells that occupy the mundane space I call my body; the other 90 percent of the cells are filled with the genomes of bacteria, fungi, protists and such, some of

---

56 M Merleau-Ponty, *The Visible and Invisible* (Evanston IL, Northwestern University Press 1968).

57 Adams (n 51).

58 Merleau-Ponty (n 52) at 4.

59 R Dolphijn and I van der Tuin, *New Materialism: Interviews and Cartographies* (Ann Arbour, University of Michigan Press 2012) at 86.

60 Philippopoulos-Mihalopoulos (n 44), ch. 2.

> which play a symphony necessary to my being alive at all … to be one is always to become with many.[61]

Haraway's 'world-making entanglements' and 'material-semiotic nodes or knots in which diverse bodies and meanings coshape one another'[62] present a fundamental generativity, a multi-scalar fecund intensiveness. Haraway's framing demands appreciation of the fact that if our ontological accounts are to be adequate to the bio-social incidents of our situated existence, they need explicitly to embrace dynamic flows and active knottings of all kinds and to acknowledge (even to celebrate) the *factity* of 'our' co-constitution with multiple others.

While the legal implications of this are explored further below, it is worth emphasising here that this broad genus of insight, as reflected by other new materialist writings, insists that such world-making material and materio-semiotic knottings go *all the way down*: what Barad calls 'intra-being'.[63] Existence is always an entangled 'becoming'. There can be *no* pre-existent, fully formed subject – such as that presupposed by law. 'We' – *all of us* – (humans and non-humans alike) are 'constituted in intra- and interaction. The partners do not precede the meeting; species of all kinds, living and not, are consequent on a subject- and object-shaping dance of encounters'.[64]

Barad's call to 'meet the Universe halfway' issues a rich invitation to embrace similarly chiasmic, energetic foldings and knottings – and to appreciate that matter has lively agencies and generates meaning. For Barad, materiality is alive with 'agential' significance – as an entanglement of intra-actions – or a 'worlding' of 'differential patterns of mattering'.[65] This has implications for traditional tensions between nature and culture: in the words of Neimanis, Barad's intra-actionist view means that 'nature' and 'culture' 'do not just "mingle" (nature affecting culture, culture impinging on nature): rather, they constantly *world one another*'.[66] That this invokes the pressing need for new forms of human

---

[61] D Haraway, *When Species Meet* (Minnesota, University of Minnesota Press 2008) at 3–4.

[62] Ibid at 4.

[63] This term is most associated with the work of Karen Barad: K Barad, *Meeting the Universe Halfway: Quantum Physics and the Entanglement of Matter and Meaning* (Durham, Duke University Press 2006).

[64] Haraway (n 61) at 4.

[65] Barad (n 63) at 140.

[66] A Neimanis, 'Alongside the Right to Water: A Posthumanist Feminist Imaginary' (2014) 5/1 *Journal of Human Rights and the Environment* 5–24 at 17, emphasis added.

epistemic humility is abundantly implied by Barad's beautiful and provocative work.

While Merleau-Ponty's ontology brings corporeal beings into a fundamentally intimate proximity – bridging the ontological rift imposed by the construction of 'the subject' (mind) and objectified 'matter' through the 'chiasm', the dynamic inter-permeation, the 'shared world of *intermundane* space which crosses over and intertwines with that of similarly embodied human beings'[67] – new materialists explicitly remove the qualifier 'human'. New materialist accounts reflect the radical entanglement of interwoven corporeal cohorts caught up in a monistic, dynamic 'worlding'.

For the purpose of the present reflection, it should be understood, moreover, that this worlding is characterised by the quintessential affectability or openness of materiality itself. We can conceptualise this affectability as vulnerability, which for corporeal beings is an inescapable – if unevenly experienced – component of their very corporeal constitution.[68] Creaturely bodies emerge vulnerable from vulnerable bodies in which they are first formed by bodily enfoldment within the corporeality of another who precedes them. 'We' humans always emerge into life pre-situated in a rich matrix of 'others', and, initially, utterly dependent upon them. And as is implied by the movement in this text between human and non-human animality, such embodied vulnerability is also a marker of a fundamental trans-species ontic commonality – *shared* (unevenly) as a condition or quality of creaturely existence itself. In short, all corporeal beings – humans included – live in a 'condition of non-intentional, heteronomous and more or less *vulnerable* openness to the surrounding world'.[69] And this is a fact calling for constant epistemic vigilance towards patterns of injustice and the imposition of constructed forms of vulnerability loaded on top of the pre-given affectability at its core.

New materialist accounts allow us to understand corporeal inter-permeation as a given of a dynamic, vulnerable planetary ontological

---

67 Williams and Bendelow (n 48) at 53.

68 M Fineman, 'The Vulnerable Subject: Anchoring Equality in the Human Condition' (2008) 20 *Yale Journal of Law and Feminism* 1–23; AB Satz, 'Animals as Vulnerable Subjects: Beyond Interest Convergence, Hierarchy and Property' in M Fineman and A Grear, *Vulnerability: Reflections on a New Ethical Foundation for Law and Politics* (Aldershot, Ashgate 2013) 171–98.

69 V Vasterling, 'Body and Language: Butler, Merleau-Ponty and Lyotard on the Speaking Embodied Subject' (2003) 11 *International Journal of Philosophical Studies* 205–23 at 214.

structure. There is, on this account, simply *no possibility at all* of a human subject surveying the world from the panoptic heights of a Cartesian or Kantian disembodied rationalism. When all is said and done, there is only an incarnate and vulnerable knowing by bodies in constant intra-action with what 'we' humans perceive to be 'the world' and calling for an explicit embrace of the 'spatial and temporal web of interspecies dependencies'.[70] In this light, existence, it turns out, is *ecological* in a deeply immanent sense, and 'we' humans ourselves are also *ecologies*.[71]

Such realities directly challenge the modern humanist foundations of contemporary environmental law. Philippopoulos-Mihalopoulos takes posthumanist new materialism into the heart of a new environmental juridical imaginary, inviting his readers to see the 'ontological plane' as an irreducibly 'open' ecology of the 'natural, the human, the artificial, the legal, the scientific, the political, the economic and so on, all of which [co-exist] on a plane of contingency and fluid boundaries'.[72]

Such an ecological ontology functions both at the level of the *on* (being) and of the ontic (beings). It is, first and foremost, an ontology of dynamic complexity and aliveness – of *movements* and *processes*. In the light of Philippoloulos-Mihalopoulos's broadly Deleuzian emphasis on dynamism, the *on* can be understood as *a becoming* expressed in a multitude of ontic becomings – a dynamic unfolding and re-folding of movements and relations in a ceaseless intensively fecund circuit: what *is*, it turns out, is a cascading, porous *multi-poesis*[73] unfolding across multiple planes.

The implications of new materialist accounts re-imagining human subjectivity are pivotal to the search for a new juridical imaginary, for, as noted above, there is an urgent need to step away from the ontological imaginary of the separative, atomistic self so implicated in the genesis of climate crisis and foundational to legal humanity as currently configured. Minimally, I suggest that the broad ontological alternative outlined here allows an appreciation of the idea that subjectivities are always, and

---

70 Haraway (n 61) at 11.

71 Again, see Haraway (n 61) at 3–4.

72 Philippopoulos-Mihalopoulos (n 44) at 2.

73 'Autopoiesis' suggests 'self-production' ('auto' for 'self'; 'poiesis' for 'production'). The term multi-poiesis is my term, and is intended to allude not only to the multiplicity of auto-poietic systems, but also to the fundamental sense in which the entire ontological plane (and earth system) can be understood as a cascading multiplicity and as a flux of dynamic multiplicities in process. It drives at the fecundity and complexity of the energetic emergences characterising the 'world'.

everywhere, an 'I'-in-the-making – never *singular* yet always *unique* – a manifestation of unique embodied *emergence* multi-poetically formed in a tumult of socio-material, natural-cultural dynamics. However, and importantly, because of the inexorable play of human power relations, this 'I' is also a 'self' agonistically constructed under conditions of conflict and injustice – as is the 'world' that we experience and perceive. Radical intra-relationality cuts all ways – and is not always benign.

In the light of all this, the meaning of the *human* – including the human of human rights and of environmental law – remains an open question. It is, in a very real sense, *undecided*. How then, are 'we' to negotiate a movement towards re-imagining the very meaning of 'we'? For reflections upon this important preliminary concern we must now turn to the project of ecological subjectivity in the making and to the development of an ecological epistemology – particularly as explored in the work of Code.

### 2.1.2 Epistemology

Code's ecological epistemology presents an important quest for the meaning of a fully ecological subjectivity, and is exactly the kind of critical philosophical reformulation required by the search for a move beyond the modern 'subject' towards renewing forms of subjectivity. Code argues that just as the traditionally dominant ontology of the West needs dethroning in favour of a more radical, open ontology, so the 'epistemic monolingualism'[74] of traditional Western epistemology needs replacement by a dynamic ecological epistemology embracing diverse ways of knowing. Given that the meaning of the 'human' remains an open question, the importance of practices of knowing and the construction of 'knowledges' move centre stage – an insight with profound implications for law as a dominant site of epistemic (hence materio-semiotic) power.

Code opens her book, *Ecological Thinking* (2006) with a statement revealing the exclusory tendencies of the ontological suppositions of the Cartesian and Kantian philosophy so foundational to the dominant socio-juridical imaginary. In particular, Code links the philosophy of Kant to the related *epistemic* hegemony undergirding the instrumentalism and reductionism characterising Western patterns of dominance and oppression. Her opening words concerning the subject of Kantian philosophy also underline the sense in which the climate crisis, as noted above, is also thoroughly legible as *a crisis of human hierarchy* mediated

---

[74] Code (n 27).

through the traditional Western construction of 'the (human) subject'. For Code, 'the Kantian moment' was decisive for what she calls the 'epistemology of mastery':

> In this book, I propose that *ecological thinking* can effect a revolution in philosophy comparable to Kant's Copernican revolution, which radically reconfigured western thought by moving 'man' to the center of the philosophical-conceptual universe. Its effects were constitutive of the humanistic, post-Enlightenment world that members of present-day affluent western societies have known. Yet Kantian philosophy was parochial in the conception of 'man' on which it turned: a recognition central to feminist, socialist, postcolonial and critical race theory, among other theoretical stances ... In placing *man* at the center of the universe ... [Kantian philosophy] tacitly promoted a picture of the world, both physical and human, that privileged and was subservient to a small class and race of people whose sex required no mention because it was presumptively male and in any case irrelevant and who were uniformly capable of achieving a narrowly conceived standard of rationality, citizenship and morality.[75]

While of course, issue can be taken with the idea that Kant's was the first philosophy to place 'man' at the centre of its discursive frame, this moment – so richly symbolic of the epistemic mastery forming the key target of Code's work – reflects, nonetheless, a high point of humanism. Let us look more closely at Code's epistemological proposal.

Code's ecological epistemology is an open one. In her words, it 'emerges from and addresses so many interwoven and sometimes contradictory issues... that its implications require multifaceted chartings'.[76] It is characterised, first and foremost, by responsible epistemic practices particularly sensitive to local, situated diversities and – importantly – 'proposes a way of engaging – if not all at once – with the implications of patterns, places and the interconnections of lives and events in and across the human and nonhuman world ... in projects of inquiry ... where epistemic and ethical-political concerns are reciprocally informative'.[77]

It is clear that Code's is an epistemology of humility entirely consistent with the claim that ecological subjectivity is a subjectivity *in the making*. She invites a deliberate and reflexive openness to a wide range of epistemic positions and modes of communicative interaction. Code's epistemic proposal is also richly body-sensitive in its willingness to

---

75 Ibid at 1.
76 Ibid at 4.
77 Ibid.

'hear' because it invites a rich, responsibilist, sensitive engagement with the materialities of 'location' – with the 'suchness' of the places and spaces in (and through) which we *are* in the world.[78]

When we bring Code's ecological epistemology into contact with the entanglements discussed above, we see precisely how human epistemic practices might be deliberately opened to a wider range of communicative interactions including – not just human speech – but action and habitus: in short, to a range of communicative openings emerging from embodied, *lived* interactions and situations – and, again, not just 'ours'. Epistemic practices, in short, may be rendered both more responsive and responsible by 'hearing' from the patterned movements and stoppings of multiple interwoven 'bodies' (of all kinds).[79]

In short, Code's ecological epistemology maps without resistance onto the ontological proposals discussed above, and while Code clearly conceives of her project as an epistemological one, her project nonetheless shares a sense of the world-making entanglements of material agency found at the heart of the work of Barad and Haraway. Indeed, Barad's insistence that ontology and epistemology stand separated by binary subject-object reductionism, and her explicit preference for the term 'onto-epistemology' has considerable resonance with the materio-semiotic depth of Code's account. Code calls for an ecology as:

> a study of habitats both physical and social where people endeavour to live well together; of ways of knowing that foster or thwart such living; and thus of the ethos and habitus enacted in the knowledge and actions, customs, social structures and creative-regulative principles by which people strive or fail to achieve this multiply realisable end.[80]

And while her language here seems *prima facie* to point towards the human, and to imply a separation between the 'physical' and the 'social', it would be a mistake, I suggest, to read her this way. It is significant that Code draws on Deleuze to fill out the notion of 'ethos' – noting that for Deleuze, *ethology* refers to 'the capacities for affecting and being affected that characterize each thing … [it studies] the compositions of relations or capacities between different things … It is … a matter of sociabilities and communities'.[81] These sociabilities/communities, of course, are also more-than-and-other-than 'human': indeed, Code's

---

78 Ibid at 5.

79 Neimanis (n 66).

80 Code (n 27) at 25.

81 G Deleuze, *Spinoza: Practical Philosophy* (San Francisco, City Lights 1988) at 125–26, cited by Code (n 27) at 26.

deployment of a Deleuzian ethology points indirectly towards the 'ethesiological'[82] yearning of the body, to the multiple relations of vulnerable/open corporealities caught up in new materialist entanglements and 'worldings'. Code's radical epistemic openness to the implications of a wide range of ethological encounters brings her project into intimate convergence with accounts such as Barad's in key respects, and Code's ecological epistemology implies – not least through its invocation of Deleuzian *ethology* – a central place for radical epistemic sensitivity to a wide range of inter- and intra-communicative materio-semiotic encounters emerging from a fully ecological ontology. In short, Code's approach can be read, without too much distortion, as a kind of onto-epistemological project.

Such accounts richly suggest a re-imagined juridical engagement with 'the world'. Quite apart from the implications of such accounts for how we might re-imagine legal process values themselves (of which more below), I have in mind, at this point, the kind of normativity emerging from epistemic openness to the materio-semiotic patterns of inter-species encounter between sheep and humans on upland commons in the United Kingdom (UK). There, bodily habits and repetitions, practices, movements, modes of stopping and of dwelling are used to guide property relations and to suggest new ways of seeing property as being, in the words of Pieraccini, the 'contingent product of humans and non-human animals'.[83] Pieraccini has in mind an appreciation of the agency of sheep in a 'mutual and dynamic crafting of people and environments'.[84] The particular context is provided by the hefts and sheep walks of upland commons in the UK – a 'heft being a unit of land to which sheep are attached. Sheep learn to recognize their heft as home and are unlikely to stray unless the overall hefting system is jeopardized'.[85] The agency of the sheep, in this story, is pivotal for the production of human proprietary relations with the land, for the normativity of these relations arises directly from the patterns, the vectors and drifts, of the sheep themselves.

---

[82] Adams (n 51).

[83] M Pieraccini, 'Property Pluralism and the Partial Reflexivity of Conservation Law: The Case of Upland Commons in England and Wales' (2012) 3 *Journal of Human Rights and the Environment* 273–87 at 273.

[84] Ibid at 280.

[85] Ibid at 281.

Pieraccini's account has considerable resonance with those of Philippopoulos-Mihalopoulos[86] and Haraway,[87] who likewise chart ethological interplays and co-formations between humans and non-human animals and others in the eco- and etho-materiality of posthuman relationalities. The work of Pieraccini and Philippopoulos-Mihalopoulos in particular, point the way towards renewing juridical conceptions of agency. Thus, Code, Pieraccini, Philippopoulos-Mihalopoulos, Barad, Haraway and Merleau-Ponty, despite theoretical differences and variant nuances in their conceptions and approaches, invite responsiveness towards a world alive with signification, in which lively matter, chiasmic movements and multiple forms of intra-action inform an insistence that there *is* no secure epistemic subject surveying an objectified world, demanding new forms of situated epistemic humility and openness. There can be no central epistemic subject because 'subject' and 'object' are now seen as contingent – the result, as Barad would argue, of a contingent, context-dependent 'cut' serving a particular, context-relative purpose but founded upon no impossible, abstract, pre-given order of fixed identity.

The implications of such accounts could hardly be more challenging to the foundations of the dominant liberal juridical imaginary. This rich, communicative meaning-making world arises as a living field of interactive intelligences. Far from being a mere inert backdrop upon which human epistemic mastery is enacted, materiality is an 'aliveness' rich with signification. This understanding suggests the necessity of transcending law's epistemic mono-culturalism to allow (and *by allowing*) non-human intelligences and agencies their place in the formation of law. Law stands called by such accounts to embrace precisely the kind of epistemic humility demanded by the fact that the 'world' is *multi-poetically* emergent, and simultaneously calls for a human juridical ecological subjectivity (or subjectivities) thoroughly responsive to the interconnectedness of multiple modes and ways of being and knowing. Simultaneously, the legal system can be re-imagined as a broader juridical ecology responsive to the emergence of norms from the midst of multiple modes of being and knowing. Such an epistemic system will be open, additionally, to what some scholars of indigenous epistemologies call 'percept ambiguity': '[o]pen-ended thought categories' characterised by implication, mystery and indeterminacy.[88]

---

86 A Philippopoulos-Mihalopoulos, 'The Triveneto Transhumance: Law, Land, Movement' (2012) 3 *Politica and Società* 447–68.

87 Haraway (n 61).

88 DM Smith, 'An Athapaskan Way of Knowing: Chipewyan Ontology' (1998) 25 *American Ethnologist* 412–32 at 412.

Finally, such an account suggests a renewing form of ethics responsive to both situatedness and to contingency – emerging from *in-the-midst-of*, rather than from an assumed/ideologically constructed epistemic centre – for as Code rightly insists, 'the center cannot hold'.[89]

### 2.1.3 Ethics

A key implication of the account traced out above is the need for ethics to be ecological in the sense proposed by Code's epistemology, which invites an ethics responsive to ethesiological bodies that spontaneously reach for relations beyond themselves;[90] to ethologically-formed (onto-)epistemologies – and highly responsive to multiple situated knowledges. Ethics, in short, gains a host of freshly awoken receptor sites sensitive to a broadening range of normative impulses emerging from a widening range of inputs, emerging from *the midst of* the situations into which corporeal beings are 'thrown' (to invoke the Heideggerian term).[91] Such ethics will also more than likely be consciously responsive to the vulnerability characterising socio-embodied experiences of the world – to the raw vulnerability of an open ontology[92] in all its unevenness and commonality: thoroughly receptive, responsibilist and humble.

Additionally, and importantly in the light of contexts in which human rights and environmental law and governance are called to work, such a renewed ethics will also need to be responsive to our growing awareness of the inseparable relationship between the most global and the most micro-situated of concerns. In particular, the deepening vulnerability characterising the material conditions of life under advanced neoliberal globalisation[93] and climate peril[94] requires a profound responsiveness to both precarity and power.

The contemporary situation is marked by profound patterns of hierarchical inequality: not simply by deepening environmental crises but by destabilising mass human displacements and crises marked by spiralling socio-economic inequalities and by the perpetration of injustice on an

---

89 Code (n 27) at 29.

90 Adams (n 51).

91 M Heidegger, *Being and Time* (Oxford, Blackwell 1962).

92 A Philippopolous-Mihalopoulos, '"... The Sound of a Breaking String": Critical Environmental Law and Ontological Vulnerability' (2011) 2/1 *Journal of Human Rights and the Environment* 5–22.

93 P Kirby, *Vulnerability and Violence: The Impact of Globalisation* (London and Ann Arbor, Pluto 2006).

94 Curry (n 28).

unprecedented scale – including forms of environmental racism,[95] socio-economic apartheid and violations enacted by deadly corporate bio-power.[96] It is marked by intense forms of destructiveness enacted upon human and non-human animal populations – especially by the imposition of unconscionable levels of suffering both within the machinery of industrial corporate production[97] and as a result of its toxic fallouts.[98] Ethics must face up to the *ethical* demands presented by living materiality itself in this situation of globalisation's stark inequalities and the socio-material vulnerabilities presented by globalising processes[99] – and a vulnerable living order in crisis.[100]

Ethics, in short, once it assumes *embodied eco-subjectivity*, can become more deeply responsive to the 'condition of non-intentional heteronomous … openness'[101] characterising, not only the human condition but the fundamental trans-species ontic commonality expressed in the quintessential *affectability* of creaturely existence itself. In short, such an ethics presents a compelling case for placing, in Fineman's words, 'actual lived experience and the human condition … at the center of our political and theoretical endeavours'[102] – but *not just the human* condition – the condition of all 'interwoven' beings and systems.

Finally, ethical reflection upon a vulnerable eco-subjectivity points towards an enduring and universal sense in which all earthling beings are *situated* – and how that might count in ethical reasoning. To return to a theme implicit in the discussion of Merleau-Ponty's work above, we can speak of an ethics and an ethically inspired politics of corporeality that speaks from *positionality*. The body always has a position. This position is not fixed or static – indeed, living bodies are a form of spatial and

---

95 L Westra and BE Lawson, *Faces of Environmental Racism: Confronting Issues of Global Justice* (Lanham MD, Rowman and Littlefield 2001).

96 K Nkrumah, *Neo-Colonialism: The Last Stage of Imperialism* (London, Butterworths 1965); U Baxi, *The Future of Human Rights* (Oxford, OUP 2001 and 2006).

97 Nibert (n 41).

98 F Pearce and S Tombs, *Toxic Capitalism: Corporate Crime and the Chemical Industry* (Aldershot, Dartmouth 1998); U Baxi, 'Writing About Impunity and Environment: The "Silver Jubilee" of the Bhopal Catastrophe' (2010) 1/1 *Journal of Human Rights and the Environment* 23–44.

99 Kirby (n 93).

100 A Grear, 'The Vulnerable Living Order: Human Rights and the Environment in a Critical and Philosophical Perspective' (2011) 2/1 *Journal of Human Rights and the Environment* 23–44.

101 Vasterling (n 69) at 214.

102 Fineman (n 68) at 2.

temporal *aliveness* that is never still (even when holding a position in a space). However, embodied beings are always characterised by a corporeality that implies, necessarily, the quality of *positionality*. This may be an unfixed positionality but the human body can never escape it to become the Cartesian eye from nowhere and everywhere. Our human ontological structure necessarily *locates* us, no matter how fleetingly or uneasily or contingently, *somewhere which is not elsewhere*. Being embodied means that humans necessarily look from *a direction – a viewpoint*. This remains true even if 'we' imagine ourselves being 'thrown in the middle' of the world[103] rather than controlling an impossible epistemic centre.[104] This fact that we always look from *somewhere* has *ethical* implications. Positionality emphasises the fact that the vulnerability of the earthling (howsoever constituted, understood or constructed) is *uneven*, radically *particular* and differentially experienced, reflecting, in a suitably paradoxical sense, the radical universal particularity of *corporeality itself*.[105] And, importantly, *subject-positionality* emerges as a central concern of how we structure *ethical encounter itself*.

## 2.2 Returning to Law

### 2.2.1 Legal subjectivity

How then, do these reflections relate to the important question of the constitution of juridical subjectivity – or to the closely related question of who we think 'we' are *in law*? As intimated above, law and legal theory have long downplayed the role of embodiment,[106] adopting a fundamentally disembodied conception of the rational actor descended from the body-excising ontological commitments of Descartes and Kant.[107] These commitments systematically undermine respect for both non-dominant human beings (and their varied ways of knowing and being) and for the living world.[108] Perhaps even more relevantly for those concerned to understand the role of law in environmental governance, the ontological

---

103 Philippopoulos-Mihalopoulos (n 44).
104 Code (n 27).
105 Grear (n 54) ch 9.
106 Bainham, Sclater and Richards (n 46) at 1.
107 Halewood (n 29).
108 L Tuhiwei Smith, *Decolonizing Methodologies: Research and Indigenous Peoples* (London, Zed Books 1999); Code (n 27).

and epistemological suppositions of law have a *profoundly* progress-inhibiting impact upon environmental law and governance regimes.[109] In the light of law's counter-productivity in this regard, now, more than ever, the selective *juridical* excision of embodiment/living materiality and its implications requires a conscious reversal. It is now urgent for a vigorous 'politics of disembodiment/embodiment'[110] to provoke law to be more responsive to the vulnerable, complex, dynamic, mutable aliveness woven through and beneath legal categories and constructs.

An urgent challenge now facing law and legal theory thus concerns precisely *how* to enable juridical categories and structures to respond to a radical ontology of openness, vulnerability and embodiment and *how* to facilitate new forms of epistemic humility in the place of the epistemology of mastery that has shaped Western liberal law and legal processes.

The challenge is profound. Liberal law is thoroughly implicated in the continuing power of disembodied ontology and its epistemology of mastery. Moreover, these commitments are thoroughly implicated in the *juridical* facilitation of a highly predatory global order relentlessly favouring, through its international *legal* structures and global regulatory institutions, the violence – both discursive and material – of Western capitalist expansionism.[111] Such well-exposed (at least by critical scholarship) impulses are intensified by neo-colonial global corporate hegemony as the old logic of colonial appropriation reaches its apotheosis – demonstrating scant regard for life in any form.[112] Indeed – it is not going too far to claim that *all life on earth* now stands thoroughly exposed to the predations and complexities of a *distinctly uneven globalised order*[113] in which law and regulatory institutions are thoroughly *productive*.[114]

---

[109] Bosselmann (n 20).

[110] U Baxi, in Grear (n 54) at xvi.

[111] S Marks, 'Empire's Law' (2003) 10 *Indiana Journal of Global Legal Studies* 449–65; S Gill, 'Globalisation, Market Civilisation and Disciplinary Neoliberalism' (1995) 24 *Millennium Journal of International Studies* 399–423; S Gill, 'Constitutionalizing Inequality and the Clash of Globalizations' (2002) 4 *International Studies Review* 47–65; A Anghie et al., *The Third World and International Order: Law, Politics and Globalization* (Boston, Martinus Nijhoff 2003).

[112] Nkrumah (n 96); Pearce and Tombs, (n 98).

[113] Radhakrishnan (n 37).

[114] F Smith, 'Indigenous Farmer's Rights, International Agricultural Trade and the WTO' (2011) 2/2 *Journal of Human Rights and the Environment* 157–77; Gill (n 111); T Evans and A Ayers, 'In the Service of Power: The Global Political Economy of Citizenship and Human Rights' (2006) 10 *Citizenship Studies*

It is not difficult to see the continuing appeal of a disembodied juridical ontology and an epistemology of 'knowing best' from an elite, plutocratic 'centre' exerting its power over the 'rest'. The contemporary juridical global order is an order of power over life and death itself – and increasingly, the corporation-dominated juridical centre of the current international order multiplies modes and sites of necropolitical pressure on the living through the power of relentless accumulation,[115] destructive exploitation, deepening socio-economic marginalisation, vicious material privation and radical insecurity for an expanding mass of people. Law's disembodied ontology and a legal epistemology forged by disembodied, ocular-centric rationalism are the perfect accompaniments to the facilitation of the immense advantages accruing to the capitalist corporation through the mechanism of legal personification and the liability shield of the corporate veil.[116] Indeed, the globalised neoliberal order *extends* the implications of personification,[117] which in its transnational manifestations now exploits protean powers of disembodied mutation to evade liability for multiple and even irreversible harms enacted against life itself.[118]

The question of entrenched corporate juridical advantage – and its relationship with processes and politics of disembodiment – presents a critical backdrop against which to locate reflection upon the need for a renewing socio-juridical imaginary within which to re-imagine the relationship between human rights and the environment. It is important to recall that the transnational corporation's uncanny powers of evasion and mutation rest, in significant part, upon *juridical characteristics* constructed as intrinsic to its current form as a legal person. Thus, while the globalised world remains full of policed borders for vulnerable embodied beings – including those human beings driven to migrate by climate change, violence, socio-economic privation and other exigencies – it remains fundamentally borderless for disembodied formations and flows

---

239–308; R Kapur, 'Human Rights in the Twenty-first Century: Take a Walk on the Dark Side' (2006) 28 *Sydney Law Review* 665–68.

[115] M Neocleous, 'International Law as Primitive Accumulation: Or the Secret of Systematic Colonization' (2012) 23/4 *European Journal of International Law* 941–62.

[116] See, J Dine, 'Jurisdictional Arbitrage by Multinational Companies: A National Law Solution?' (2012) 3/1 *Journal of Human Rights and the Environment* 44–69.

[117] M Neocleous, 'Staging Power: Marx, Hobbes and the Personification of Capital' (2003) 14 *Law and Critique* 147–65.

[118] Dine (n 116).

of capital[119] that retain potent ideological and practical advantages arising from the special form of corporate juridical subjectivity.

There is a need for *vigilance* concerning the constitution of juridical subjectivity. If, as I have argued, our human situation demands a fundamental reformulation of our very sense of who 'we' are, how 'we' know and what 'our' responsibility 'in the world' looks like, then our theorisation and agenda for doctrinal reformulation must *necessarily* revisit the question of law's subject or person – a point rendered even more powerful in the light of the unevenness between corporate legal subjects and other subjects of law – including billions of human beings.

The question of how to re-imagine the subjects of law now seems unavoidable – yet it is a question almost routinely ignored by the courts (which award legal subjectivity, frequently without adequate normative enquiry),[120] by most doctrinal legal theory,[121] and by campaigners – whether anti-capitalist[122] or pro-animal rights and the like. In the light of rights for nature arguments and related developments, there is an intense irony in the gap now attending the question of law's subject[123] in an age of ecological crisis.

As is well known, current legal approaches tend to leave the fundamental suppositions informing the construction of human subjectivity in relation to 'the environment' largely untouched. Indeed, it can be argued that existing strategies such as ecosystems approaches, in reality, merely attempt to minimise damage to natural ecosystems while facilitating, as far as is possible, the priorities of capitalism, and that various 'eco-centric' initiatives within the existing international environmental law and

---

119 Baxi (n 96) (2006).

120 See J Finnis, 'The Priority of Persons' in J Horder, *Oxford Essays in Jurisprudence: Volume IV* (Oxford, OUP 2000); M Emberland, *The Human Rights of Companies: Exploring the Structure of ECHR Protection* (Oxford, OUP 2006); Grear (n 54).

121 Harvard note: 'Notes: What We Talk About When We Talk About Persons: The Language of a Legal Fiction' (2001) 114 *Harvard Law Review* 1746; N Naffine, 'Who are Law's Persons? From Cheshire Cats to Responsible Subjects' (2003) 66 *Modern Law Review* 356–67; J Berg, 'Of Elephants and Embryos: A Proposed Framework for Legal Personshood' (2007) 59 *Hastings Law Journal* 369–96.

122 Neocleous (n 117).

123 Berg (n 121).

governance system are entirely continuous with capitalist strategies of expansionism.[124]

Philippopoulos-Mihalopoulos suggests the need for a new, critical environmental law (CEL) resisting the dialectics of 'anthropocentric/ecocentric'. He argues that Barry Commoner's first law of ecology (that 'everything is connected to everything else')[125] is *best realized* in an ecology that is '*processual*' rather than through *values*.[126] Such an approach, based as it is upon a broadly Deleuzian foundation, focuses upon what it is that concepts *do*, rather than what they *are*. The processual ecology of CEL turns us towards an ontology of movement, to 'various elements that repeat themselves in nature and humanity in the form of *processes/products*' in an *open ecology* that 'combines the natural, the human, the artificial, the legal, the scientific, the political, the economic and so on, all of which co-exist … on a plane of contingency and fluid boundaries'.[127] On such an account, *all* categorizations become *contingent*. The focus is placed upon the constant circulation and ontological flux in which human beings as cognitive-corporeal beings (to use Code's descriptor) are inescapably engaged *as* living systems *within* living systems always becoming one *with* and *as* 'many' (to evoke Haraway's insight). CEL responds to the tired closures of law's Cartesian and Kantian assumptions by proposing a new, open ecology of knowledges in a plane of dynamic uncertainty. Law is pressed away from its carefully defended epistemic centre and invited to rely upon a multiplicity of insights from which choice – in full knowledge of the contingency of choices – factors in processes and strategies reflexively sensitive to the interplay of vulnerabilities, capacities and movements characterising life on Earth *as far as we understand them for now and even then, incompletely*.

The epistemic centrality of contingency is thoroughly implied by such an approach and invites, directly, the kinds of responsible epistemic engagements at the heart of Code's call for ecological thinking. The subjectivity implied by CEL is a subjectivity of in-the-midst-of (rather than of the knowing subject as the 'centre') – a subjectivity inviting an epistemology of openness – of humility – of a willingness to 'hear' from

---

124 See e.g., S Bohm, MC Misoczky and S Moog, 'Greening Capitalism? A Marxist Critique of Carbon Markets' (2012) 33 *Organization Studies* 1617–38, and the references cited therein.

125 B Commoner, *The Closing Circle: Nature, Man and Technology* (New York, Knopf 1971) at 7.

126 Philippoloulos-Mihalopoulos (n 44) at 2, emphasis added.

127 Ibid at 2; see also Philippopoulos-Mihalopoulos (n 86).

multiple situated beings and constituencies. This, I suggest, implies certain key juridical mechanisms as a form of responsiveness: a processual ontology suggests a processual epistemology pointing towards the importance of processual mechanisms of communication – both formal and informal.

Minimally, in this light, the centrality of juridical participatory processes becomes a fully ecological concern, pointing (as Code would surely agree) towards transformed advocacy practices – and as Stone would argue – towards the widening of rules on standing.[128] It points beyond this to the juridical empowerment of informal participatory processes – to the vital role of constructing diverse *fora* of situated encounter in which diverse insights, interests and vulnerabilities are brought into productive negotiation (albeit contingently). Such negotiation, in other words, could allow law and legal processes to respond more richly to situated subjectivities – providing a juridical deployment of a fully ecological epistemology in Code's rich sense.

And even within traditional formal legal processes it might be possible to expand the horizons of reference. The ecological meanings of *audi alteram partem*, for example, could be explored – perhaps opening the principle's inference to involve a far wider range of modes of listening and hearing from as yet unconsidered constituencies of 'speech'. Quite apart from the obvious need to hear from a range of indigenous traditions and other situated communities of knowledge, legal processes could also 'hear from' wider constituencies of being – not just by opening up the rules of standing but by broadening the law's epistemic conception of the interactions between beings and systems. The central point is that legal processes can become more sensitive to the multiple implications of the materio-semiotic knottings of Haraway's rich invitation to 'meet', as it were, between species of all kinds. Juridical processes could facilitate and inform this meeting by allowing for the kind of situated norm-generation implied by Pieraccini and Philippopoulos-Mihalopoulos and their accounts of interwoven humans and sheep. The epistemic and normative significance of a wide range of intra-actions could be explored and embraced. Science could inform this process by directing a whole range of non-human 'inputs' into such an open juridical epistemology, producing a more eco-inclusive notion of *audi alteram partem* responsive

---

128 CD Stone, 'Should Trees Have Standing? Towards Legal Rights for Natural Objects' (1972) 45 *Southern California Law Review* 450–87.

to measurements of, for example, ecological integrity,[129] but resisting assumptions concerning the completeness or epistemic dominance of scientific knowledge itself. Such inputs might produce a reconceived '*alteram*' – the '*alter*' (or 'other') no longer being merely confined to a human or other existing legal protagonist (a company, a local authority and so forth), but being understood to embrace multiple life-forms and systems seen as *co-participants* in the formation of normative outcomes. This, arguably, requires the emergence of a renewing socio-juridical imaginary. Only then, arguably, can legal processes expand explicitly to embrace (adequately) the normative insistences of many different modes of being.

The de-centring of human mastery achieved by these moves also suggests richer potential depths to the *nemo judex in sua causa* element of legal process rules, for attention is directed away from the privileged epistemic 'centre' towards a diverse ecology of knowledges and towards broader questions of inter- and intra-species justice than law typically embraces.

These broader questions of inter- and intra-species justice could also have particularly powerful implications for the imperative of addressing the radical unevenness of injustice in a neo-liberal global order – and for de-centring, for example, the current dominance of corporations over participatory governance structures – including over environmental governance processes.[130]

A new juridical imaginary is possible – and vital to build. The philosophical foundations are being laid in new materialist and posthumanist philosophy – and in imaginative experiments in ecological epistemology. Law and legal theory, by responding to these new, passionate visions of an interwoven world, would follow – to borrow again the

---

[129] L Westra, K Bosselmann and R Westra, *Reconciling Human Existence with Ecological Integrity: Science, Ethics, Economics and Law* (Padstow, Earthscan 2008).

[130] K Bosselmann, 'Towards Multi-Level Governance for Sustainability: Trends and Challenges' in 'New Zealand and the EU: Contested Futures: Sustainability, Governance and International Human Rights' (2010) *Europe-New Zealand Research Series* 1–72; A Grear, 'Multi-Level Governance for Sustainability: Reflections from a Fractured Discourse: A Response to Bosselmann' in 'New Zealand and the EU: Contested Futures: Sustainability, Governance and International Human Rights' (2010) *Europe-New Zealand Research Series* 73–141.

words of Rhadakrishnan – a deeply ethical impulse. The time has come. New future histories summon the legal theoretical imagination – and potential new interpretations and modes of legal process augur meaningful, practical ways of responding to the call.

# Index